Cambridge Primary

Ready to Go Lessons for Science

Step-by-step lesson plans for Cambridge Primary

Stage 5

Judith Amery

Series editor: Judith Amery

HODDER EDUCATION
AN HACHETTE UK COMPANY

Note: Whilst every effort has been made to carefully check the instructions for practical work described in this book, schools should conduct their own risk assessments in accordance with local health and safety requirements.

Every effort has been made to trace all copyright holders, but if any have been inadvertently overlooked the Publishers will be pleased to make the necessary arrangements at the first opportunity.

Although every effort has been made to ensure that website addresses are correct at time of going to press, Hodder Education cannot be held responsible for the content of any website mentioned in this book. It is sometimes possible to find a relocated web page by typing in the address of the home page for a website in the URL window of your browser. Websites included in this text have not been reviewed as part of the Cambridge Endorsement Process.

Hachette UK's policy is to use papers that are natural, renewable and recyclable products and made from wood grown in sustainable forests. The logging and manufacturing processes are expected to conform to the environmental regulations of the country of origin.

Orders: please contact Bookpoint Ltd, 130 Milton Park, Abingdon, Oxon OX14 4SB. Telephone: (44) 01235 827720. Fax: (44) 01235 400454. Lines are open 9.00–5.00, Monday to Saturday, with a 24-hour message answering service. Visit our website at www.hoddereducation.com.

© Judith Amery 2013
First published in 2013 by
Hodder Education,
An Hachette UK Company
338 Euston Road
London NW1 3BH

Impression number   5 4 3 2
Year   2017 2016 2015 2014

All rights reserved. Apart from any use permitted under UK copyright law, the material in this publication is copyright and cannot be photocopied or otherwise produced in its entirety or copied onto acetate without permission. Electronic copying is not permitted. Permission is given to teachers to make limited copies of individual pages marked © Hodder & Stoughton Ltd 2013 for classroom distribution only, to students within their own school or educational institution. The material may not be copied in full, in unlimited quantities, kept on behalf of others, distributed outside the purchasing institution, copied onwards, sold to third parties, or stored for future use in a retrieval system. This permission is subject to the payment of the purchase price of the book. If you wish to use the material in any way other than as specified you must apply in writing to the Publisher at the above address.

Cover illustration by Peter Lubach
Illustrations by Planman Technologies
Typeset in ITC Stone Serif Medium 10/12.5 by Planman Technologies
Printed in Great Britain by CPI Group (UK) Ltd, Croydon, CR0 4YY

A catalogue record for this title is available from the British Library.

ISBN: 978 1444 177862

Contents

Introduction	4
Overview chart	6

Term 1

    Unit 1A: 5.1 The way we see things   8
                  Unit assessment   35

    Unit 1B: 5.2 Evaporation and condensation   37
                  Unit assessment   64

Term 2

    Unit 2A: 5.3 The life cycle of a flowering plant   66
                  Unit assessment   96

    Unit 2B: 5.4 Investigating plant growth   98
                  Unit assessment   115

Term 3

    Unit 3A: 5.5 Earth's movements   117
                  Unit assessment   153

    Unit 3B: 5.6 Shadows   155
                  Unit assessment   191

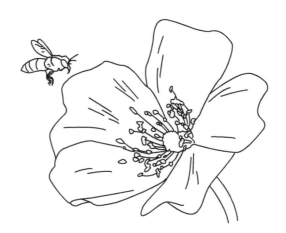

Introduction

About the series

Ready to Go Lessons is a series of photocopiable resource books providing creative teaching strategies for primary teachers. These books support the revised Cambridge Primary curriculum frameworks for English, Mathematics and Science at Stages 1–6 (ages 5–11). They have been written by experienced primary teachers to reflect the different teaching approaches recommended in the Cambridge Primary Teacher Guides. The books contain lesson plans and photocopiable support materials, with a wide range of activities and appropriate ideas for assessment and differentiation. As the books are intended for international schools we have taken care to ensure that they are culturally sensitive.

Cambridge Primary

The Cambridge Primary curriculum frameworks show schools how to develop the learners' knowledge, skills and understanding in English, Mathematics and Science. They provide a secure foundation in preparation for the Cambridge Secondary 1 (lower secondary) curriculum. The ideas in this book can also be easily incorporated into existing curriculum frameworks already in your school.

How to use this book

This book covers each of the units of the scheme of work for Science at Stage 5. It can be worked through systematically (as all the learning objectives are covered), or used to support areas where you feel you need more ideas. It is not prescriptive – it gives ideas and suggestions for you to incorporate into your own teaching as you see fit.

Each step-by-step lesson plan shows you the learning objectives you will cover, the resources you will need and how to deliver the lesson.

Each lesson includes a Starter activity, Main activities and a Plenary that draws the lesson to a close and recaps the learning objectives. Success criteria are provided in the form of questions to help you assess the learners' level of understanding. The 'Differentiation' section provides support for the less-able learners and extension ideas for the more able.

For each lesson plan there is at least one supporting photocopiable activity page. At the end of each unit there are also suggestions for assessment activities. Answers to activities can be found at www.hoddereduction.com/cambridgeextras.

Learning objectives

The *Science Curriculum Framework* provides a set of learning objectives for each stage. At the start of each lesson you need to re-phrase the learning objectives into child-friendly language so that you can share them with the learners at the outset. It sometimes helps to express them as *We are learning to / about ...* statements. This really does help the learners to focus on the lesson's outcomes. For example: 'Know that water is taken in through the roots and transported through the stem' (Stage 3) could be introduced to the learners at the start of the lesson as: *We are learning about the journey water takes through a plant*. To avoid unnecessary repetition we have not included such statements at the start of each lesson plan but it is understood that the teacher would do this.

The overview chart on pages 6–7 shows you how the learning objectives are covered in the lessons in this book.

Time commitment

Teachers should be aware that the recommended time commitment for Science at Stages 1 and 2 is an hour to one and a half hours per week. This could be as a whole afternoon or two or three shorter sessions, depending on timetabling arrangements in your school. The recommended time commitment at Stages 3 to 6 is at least two hours per week. This provides ample time to carry out practical work. Again, it can be timetabled as one long or several shorter sessions. We have, however, provided the same number of lesson plans for you for all six stages to provide choice and variety. Please select the most appropriate lessons for your class for Stage 5 to suit the amount of time available to you.

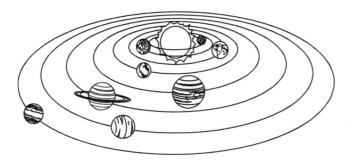

Success criteria

These are the measures that the teacher and, eventually, the learner will be able to use to assess the outcome of the learning that has taken place in each lesson. They are included as a series of questions, which will help you as teacher to assess the learners' understanding of the skills and knowledge covered in the lesson.

Scientific enquiry skills

Science teaching is concerned with more than just the learning of scientific facts. Scientific enquiry skills are also **essential**.

The activities in these books will show you how to incorporate scientific enquiry skills in order to link practical skills alongside thinking skills using the Cambridge Primary Science Programme. Scientific enquiry is embedded in the curriculum in the Biology, Chemistry and Physics strands. The skills of scientific enquiry are on-going in each stage and between stages. These skills need to be used regularly, in familiar and new contexts, in order for the learners to become young scientists who are capable of questioning, reasoning and finding answers through scientific investigation. Every lesson in this book has links to at least one scientific enquiry learning objective.

The key to successful scientific enquiry teaching lies in providing the learners with opportunities to learn by doing, that is, through **active learning**.

Formative assessment

Formative assessment is on-going assessment that occurs in every lesson and informs the teacher and learners of the progress they are making, linked to the success criteria. The types of questions to ask that will support teachers in making formative assessments have been incorporated into each lesson in the 'Success criteria' sections.

One of the advantages of formative assessment is that any problems that arise during the lesson can be responded to immediately. Formative assessment influences the next steps in learning and may influence changes in planning and / or delivery for subsequent lessons.

Summative assessment

Summative assessment is essential at the end of each unit of work to assess exactly what the learners know, understand and can do. The assessment sections at the end of each unit are designed to provide you with a variety of opportunities to check the learners' understanding of the unit. These activities can include specific questions for teachers to ask, activities for the learners to carry out (independently, in pairs or in groups) or written assessment.

The information gained from both the formative and summative assessment ideas can then be used to inform future planning in order to close any gaps in the learners' understanding as recommended by *Assessment for Learning* (AFL).

Safety

All the lessons in this book have been written with safety in mind. However, please ensure that you are aware of and conform to any national, regional or school regulations for safety as you conduct any of the activities in this book. Always be aware of skin and food allergies or intolerances and obtain parental consent for the learners to participate in tasting activities. If necessary, make sure that you undertake a risk assessment of potential hazards before undertaking activities. It is important to ensure that the learners are aware of safety considerations when carrying out practical activities.

Appropriate use of ICT

At the planning stage teachers need to consider how the use of ICT in a lesson will enhance the learning process. Ensure that the ICT resources you use support and promote the learners' understanding of the learning objectives. Activities included in this book have been designed to be carried out without the need for state-of-the-art ICT facilities. Suggestions have also been included for schools with internet access and / or the use of interactive whiteboards. This is in order to cater for most teachers' needs.

In these lessons the author sometimes asks for the teacher to display an enlarged version of the photocopiable page at the front of the class. We have not specified whether this should be using an overhead projector, interactive whiteboard or flipchart, as schools will have different resources available to them.

We hope that using these resources will give you confidence and creative ideas in delivering the Cambridge Primary curriculum framework.

Judith Amery, Series Editor

Overview chart

	Lesson	Framework codes	Page
Term 1 — **Unit 1A: 5.1 The way we see things**	Light sources 1	5Pl6 5Ep2 5Eo1	8
	How do we see?	5Ep2 5Ep5 5Pl6	11
	The eye	5Ep2 5Eo1 5Pl6	13
	Reflections	5Ep2 5Ep5 5Pl7	16
	Reflective surfaces	5Ep4 5Eo8 5Pl7	18
	Making periscopes	5Ep1 5Ep5 5Pl7	23
	Light beams	5Ep3 5Pl8	25
	How many images?	5Eo4 5Eo7 5Pl7	27
	Comparing reflections	5Eo1 5Pl8	29
	Optical illusions	5Ep1 5Pl6	32
Unit assessment			35
Unit 1B: 5.2 Evaporation and condensation	What is evaporation?	5Eo1 5Eo2 5Cs1	37
	The water cycle	5Cs1 5Cs2 5Eo2	39
	What is condensation?	5Eo1 5Cs2 5Cs3	41
	Hot or cold?	5Eo2 5Eo3 5Cs4	46
	Melting and boiling points of water	5Eo2 5Eo6 5Cs4	48
	How long can you keep an ice cube for?	5Ep3 5Ep4 5Cs4	52
	Making solutions	5Ep6 5Eo1 5Cs5	55
	Soluble or insoluble?	5Ep3 5Eo5 5Cs5	57
	Growing crystals	5Ep6 5Eo1 5Cs5	59
	What's in it?	5Ep3 5Eo5 5Cs5	61
Unit assessment			64

	Lesson	Framework codes	Page
Term 2 — **Unit 2A: 5.3 The life cycle of a flowering plant**	Flowers and fruits	5Bp2 5Eo1	66
	Seeds	5Bp2 5Eo1 5Eo4	68
	Planting seeds	5Bp2 5Ep6 5Eo3	70
	Seed dispersal	5Bp3 5Eo1	72
	Investigating seed dispersal	5Bp3 5Eo1	74
	Germination	5Bp7 5Ep3 5Ep6	77
	Evidence of germination	5Bp7 5Eo3 5Eo8	80
	Insect pollination	5Bp5 5Bp7 5Eo1	83
	Plant structure	5Bp6 5Bp2 5Eo1	86
	Plant parts in pollination	5Bp5 5Bp6 5Eo1	90
	The life cycle of a flowering plant	5Bp2 5Bp7 5Eo1	93
Unit assessment			96

	Lesson	Framework codes	Page
Unit 2B: **5.4 Investigating plant growth**	Conditions for germination	5Bp4 5Eo7	98
	What do healthy plants need?	5Bp1 5Ep2	100
	What do plants need light for?	5Bp1 5Eo1	103
	How does temperature affect plant growth?	5Eo2	106
	How do plants grow best – with or without water?	5Bp1 5Ep4 5Eo2	109
	Photosynthesis	5Bp1 5Eo1	112
Unit assessment			115

	Lesson	Framework codes	Page
Unit 3A: **5.5 Earth's movements**	Earth, Sun and Moon	5Pb1 5Pb3 5Eo1	117
	The Sun in the sky	5Pb1 5Eo1 5Eo2	120
	Day and night	5Pb1 5Pb2 5Eo1	123
	How long is a day?	5Pb2 5Eo4 5Eo7	126
	What makes a year?	5Pb3 5Eo1 5Eo4	129
	The Solar System	5Pb4 5Eo1	132
	Space scientist – Aristarchos	5Pb4 5Ep1	137
	Space scientist – Pythagoras	5Pb4 5Ep1	140
	Space scientists – Copernicus and Galileo	5Pb4 5Ep1	144
	Researching space scientists	5Pb4 5Ep1	146
Unit assessment			153
Unit 3B: **5.6 Shadows**	Light sources 2	5Pl1 5Eo1	155
	How shadows are formed	5Pl1 5Eo1	158
	What makes the sharpest shadow?	5Pl1 5Eo1 5Eo7	161
	Changing the size of shadows	5Pl2 5Eo2	163
	Investigating shadows	5Pl2 5Eo1 5Eo5	166
	What affects the size of a shadow?	5Pl2 5Ep2 5Eo1	170
	How do shadows change throughout the day?	5Pl3 5Ep2 5Eo5	173
	Shadows during the course of a day	5Pl3 5Ep2 5Eo1	177
	Measuring light intensity	5Pl4 5Eo3 5Eo8	179
	How does light travel through some materials?	5Pl5 5Ep3 5Ep6	181
	Fun with shadows!	5Pl5 5Eo1	188
Unit assessment			191

Term 3

Unit 1A: 5.1 The way we see things

Light sources 1

Learning objectives

- Know that we see light sources because light from the source enters our eyes. (5Pl6)
- Use observation and measurement to test predictions and make links. (5Ep2)
- Make relevant observations. (5Eo1)

Resources

Flipchart paper and markers; a dark box, e.g. a shoebox, painted black on the inside, with a small peep-hole at one end; a torch; a circuit with a small bulb; a mirror; a piece of jewellery containing a gemstone; aluminium foil; nightlights; matches; tray of damp sand (to place nightlights in – for safety purposes); photocopiable pages 9 and 10.

Starter

- Ask the learners to list as many light sources as they can in one minute.
- With talk partners, ask them to share their lists and agree on things that are light sources.
- In small groups, ask the learners to compile a mind map on flipchart paper, including each group member's contributions. A mind map is sometimes referred to as a concept map or a spider diagram. The words 'light sources' should be written in the centre of the paper, then lines drawn that radiate off the centre to include things associated with the title.
- Invite the learners to share their completed mind maps with the whole class. Explain that they need to describe reasons why they have included each source. Display the mind maps around the room, or compile a whole-class mind map for later reference and display.

Main activities

- In the same small groups, ask the learners to use the mind map to write lists of groups of objects that produce light in a similar way.
- Ask each group in turn to explain to the whole class about one of the groups they have made and why they chose it.
- If they include the Moon, introduce the term 'reflector' to describe things that reflect light from somewhere else. (The Moon reflects the Sun's light.)
- Explain to the learners that in this lesson they will investigate some things that can be classified as light sources.
- Give out photocopiable pages 9 and 10 and demonstrate what the learners have to do to test objects as light sources. Demonstrate using a nightlight, that is, a naked flame – do not allow the learners to test any naked flames. Describe how they should record their findings. Photocopiable page 10 offers a more structured version for learners who need support.

Plenary

- Ask the learners to bring into school examples of light sources for a classroom display. Use the display as a discussion point at the start of the Science lessons in this unit of work to remind them about light sources. Ask them to think about which would be better light sources than others.

Success criteria

Ask the learners:

- How did you decide which objects are light sources?
- Is a really shiny piece of aluminium foil a light source?
- What is a light source? (Something that gives off its own light.)
- How can we see things?
- Why can we see the light source?

Ideas for differentiation

Support: Give these learners photocopiable page 10 to complete.

Extension: Ask these learners to research natural light sources, such as the Sun or glow-worms, and produce an information leaflet about one source.

Name: _____

Sources of light

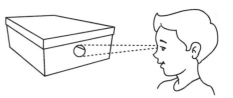

You will need:
A dark box with a peep-hole in it, a selection of possible light sources.

What to do
- Place an object in the box.
- Close the lid and look through the peep-hole.
- Can you see any light?
- Repeat using different objects.

Results (what happened)

1. Record your results in the box below for the objects you tested.

2. What is a light source?
   _____

Name: _____

Sources of light

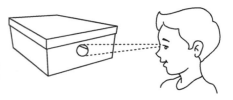

You will need:
A dark box with a peep-hole in it, a selection of possible light sources.

What to do
- Place an object in the box.
- Close the lid and look through the peep-hole.
- Can you see any light?
- Repeat using different objects.

Results
Complete the table to show what happened.

Object	Light source? ✓ or ✗

Unit 1A: 5.1 The way we see things

How do we see?

Learning objectives

- Use observation and measurement to test predictions and make links. (5Ep2)
- Collect sufficient evidence to test an idea. (5Ep5)
- Know that we see light sources because light from the source enters our eyes. (5Pl6)

Resources

Box painted black on the inside with a peep-hole at one end from the previous lesson; a range of everyday classroom objects, e.g. a book, a pair of scissors, a ruler, a hole punch, a stapler; a candle, matches, damp sand in a tray (for safety); photocopiable page 12.

Starter

- Play the game 'What is it?': Select different objects for the learners to identify, either inside the box from the previous lesson or in a darkened corner of the room. If your room cannot be made sufficiently dark, you could try using a blindfold instead (if the learners are happy to do this).
- Ask the learners to take turns to choose and select objects for other learners to identify.
- Discuss responses, asking whether the object selected was easy or difficult to identify. Ensure that the learners understand that it is easier to see things clearly in the light.

Main activities

- Explain that in this lesson the learners will find out about how we are able to see things.
- Light a candle and use it as the light source for this observation. Explain that the candle is placed in damp sand for safety so that it can be extinguished quickly if necessary. Ensure that the learners keep a safe distance away from the naked flame. Only an adult should use the matches.
- Ask the learners to discuss with talk partners how they see the candle flame.
- Listen to responses from different learners. Do not comment on their responses as they should be able to state by the end of the lesson if their original ideas were correct or not.
- Give out photocopiable page 12. Ask the learners to draw a picture to show how they think they can see the candle flame. Ask them to draw arrows on their diagram to show how the light travels to enable us to see. The arrows depict light rays. Remind the learners that light travels in straight lines.
- Compare diagrams when they are complete. Do this in pairs, or invite the learners to volunteer to share their diagram with the rest of the class. This will indicate the learners' understanding of how we see.

Plenary

- Emphasise that we see objects because light enters our eyes, not because light travels from our eyes. This is a difficult concept for many learners at this stage.
- Explain that light can be perceived by us in two different ways. First, light from a source can travel directly to our eyes, like the light from the candle flame. Second, light from a light source can bounce off an object and enter our eyes, for example if we are looking at a flower, light from the Sun (light source) can bounce off the flower and into our eyes.

Success criteria

Ask the learners:

- How does light travel?
- Why can you see the candle flame?
- How do we see objects?
- What do we need to be able to see clearly?

Ideas for differentiation

Support: Assist these learners in completing photocopiable page 12. Discuss the direction in which light travels, emphasising that it travels from the source to the eye.

Extension: Ask these learners to draw a labelled diagram on the back of photocopiable page 12 to show how they are able to read a book.

Name: _____

How do we see?

1. Draw a diagram to show how we see the light from a candle flame. Use arrows to show how the light travels.

2. Write a sentence to explain how this happens.

_____

_____

Unit 1A: 5.1 The way we see things

The eye

Learning objectives

- Use observation and measurement to test predictions and make links. (5Ep2)
- Make relevant observations. (5Eo1)
- Know that we see light sources because light from the source enters our eyes. (5Pl6)

Resources

Mirrors; coloured pencils; A4 paper; a model of the eye (if available) or a poster or diagram; photocopiable pages 14 and 15.

Starter

- Give each pair of learners a mirror. Ask them to use the mirror to look at their eyes and describe them to their talk partner.
- Ask the learners to draw a detailed (front view) picture of their eye – they should use coloured pencils and label as many parts as they can see.
- Invite the learners to share their finished eye pictures with the whole class. Encourage use of the words 'eyeball', 'eyelashes', 'eyebrows', 'pupil', 'white of the eye', 'tear duct', 'retina', 'lens'. Use this as an opportunity to find out the terminology that the learners are already familiar with. They may be familiar with all or just some of these words.

Main activities

- Explain that in this lesson the learners will learn about the internal structure of the eye and how this helps us to see.
- Use the model, if available, or poster or picture to help you discuss and describe the internal structure of the eye.
  - Describe the pupil as the black part of your eye where light enters.
  - Explain that the job of the iris is to make the pupil larger or smaller, depending on the amount of light available.
  - Tell the learners that the lens helps us to focus (like a camera lens) and that the optic nerve joins the eye to the brain.
  This is the extent of the knowledge that they need about the structure of the eye.
- Refer to photocopiable page 14 for the relevant parts that the learners need to know. Give photocopiable page 14 to the learners who need support and photocopiable page 15 to all the other learners.

Plenary

- Use the model or poster or picture and ask the learners to identify the different parts of the eye already talked about.
- Ask them to describe the function/s of each part.
- Discuss what happens to the pupil in dark and light. (Alters in size.)

Success criteria

Ask the learners:

- What nerve connects your eye to your brain?
- Where is your pupil?
- What does the pupil do?
- How does the iris help the pupil to work?
- What is the main job of the lens?

Ideas for differentiation

Support: Assist these learners in completing photocopiable page 14. Ensure that their sentences make sense.

Extension: Ask these learners to find out about how spectacles help some people to see more clearly. They could survey the class or their family members to ask why some people wear glasses, for example just for reading, that is, close-up work, or for distance.

Name: _____

The eye

1. Use the words below to label the diagram.

   | iris | lens | pupil | optic nerve |

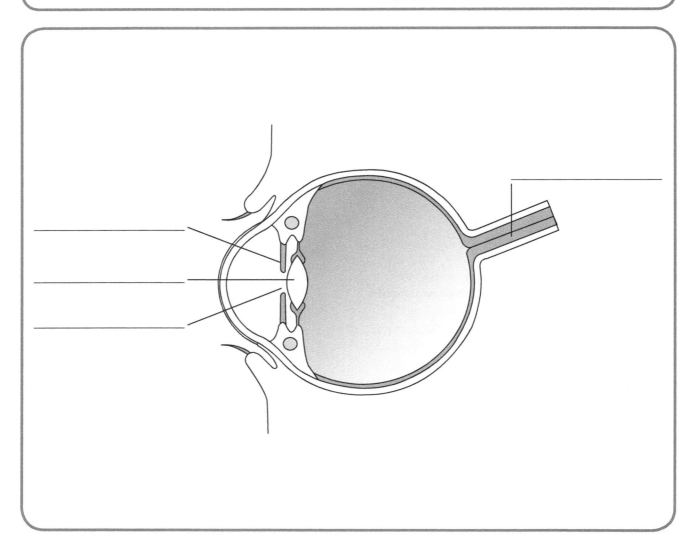

2. Draw a line to match the start and ending of the sentences below.

The iris is	helps us to focus.
The pupil is	goes to the brain.
The lens	the coloured circle.
The optic nerve	the black centre.

Name: _____

The eye

1. Label this diagram of the human eye.

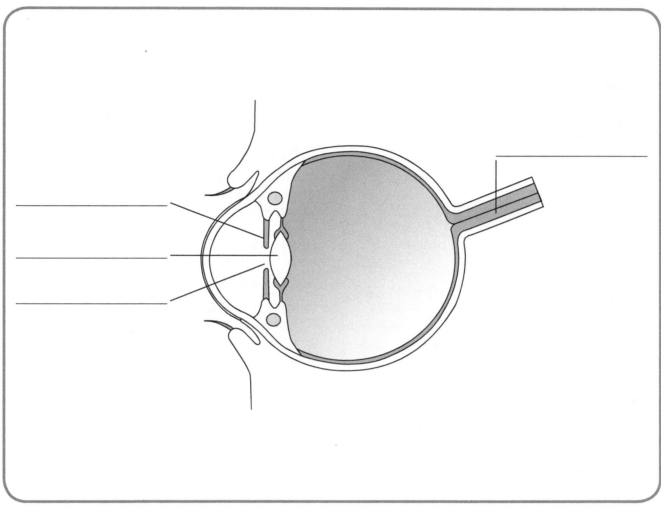

2. Write a sentence to say what each part of the eye does.

   a) _____
   _____

   b) _____
   _____

   c) _____
   _____

   d) _____
   _____

Cambridge Primary: Ready to Go Lessons for Science Stage 5 © Hodder & Stoughton Ltd 2013

Unit 1A: 5.1 The way we see things

Reflections

Learning objectives

- Use observation and measurement to test predictions and make links. (5Ep2)
- Collect sufficient evidence to test an idea. (5Ep5)
- Know that beams / rays of light can be reflected by surfaces including mirrors, and when reflected light enters our eyes we see the object. (5Pl7)

Resources

Flat (plane) mirrors; photocopiable page 17.

Starter

- Give each learner a mirror and ask them to make a happy face and then a sad face.
- What can they see happening in the mirror? Discuss their responses.
- Ask the learners to touch their right ear with their right hand, then to observe and comment on what they see with talk partners. Invite them to share their findings with the rest of the class.
- Ask the learners to prepare to wink with their left eye (check that they know right from left, first!). Then ask them to predict where they will see the reflection. Ask them to try it and see. What happens?
- Next ask the learners to hold the mirror on a piece of paper with one hand and to write their name with the other hand. What do they see? Can they think of any examples where this might be useful? (The writing on an ambulance in the UK, for example, is often written backwards so that it can be read in a car mirror.)

Main activities

- Discuss the properties of mirrors, for example they are smooth and shiny.
- Ask the learners what we call the image you can see in a mirror (your reflection).
- Collect in the mirrors.
- Give out photocopiable page 17. Explain that the learners are going to predict which capital (upper case) letters will look the same in the mirror as they do on paper. Ask them to make their predictions first.
- Give the mirrors back out again. Let the learners try out their predictions. Were they correct?

Plenary

- Go over the learners' responses to photocopiable page 17. Discuss any misunderstandings or disagreements.

Success criteria

Ask the learners:

- What does your reflection do when you pull a face in the mirror?
- What would your reflection look like if you put your right hand over your right ear?
- Show me a capital (upper case) letter that looks the same in the mirror as it does on paper.
- Show me a letter that looks different in the mirror from how it is written.

Ideas for differentiation

Support: Provide these learners with a paper copy of the alphabet written in capital (upper case) letters for reference during this task. Work alongside them in a small group, supervising and questioning them appropriately to keep them focused and on task.

Extension: Challenge these learners to write a message that can be read when viewed through a mirror. Allow them to ask other classmates to predict what the message might be before using a mirror to read and check it.

Name: _____

Reflections: making predictions

1. Which capital (upper case) letters will appear the same in a mirror when written down?

2. Write the letters that you think will look the same in the box below.

3. Now try it and see.

4. Tick (✓) those that were correct. Write any others in the box below.

5. How many did you get right? _____

Cambridge Primary: Ready to Go Lessons for Science Stage 5 © Hodder & Stoughton Ltd 2013

Unit 1A: 5.1 The way we see things

Reflective surfaces

Learning objectives

- Use knowledge and understanding to plan how to carry out a fair test. (5Ep4)
- Interpret data and think about whether it is sufficient to draw conclusions. (5Eo8)
- Know that beams / rays of light can be reflected by surfaces including mirrors, and when reflected light enters our eyes we see the object. (5Pl7)

Resources

Flipchart and markers; collection of objects and / or materials, e.g. aluminium foil, metal spoons, saucepan lids, mirrors, plastic, paper, wood, glass; torches and spare cells (batteries); photocopiable pages 19, 20, 21 and 22.

Starter

- Ask the learners to discuss with talk partners which materials reflect light well and which do not reflect light so well.
- Share ideas as a whole class.
- Compile a class list of these suggestions.

Main activities

- Explain to the learners that in this lesson they will find out which materials reflect light best. Instruct them that they will need to think about how they will be able to tell if a material reflects light or not. (Guide their thinking into using a torch and observing how its light beam is affected when it is shone on different materials. Demonstrate how a light beam is reflected when it hits a plane mirror. Encourage them to adopt a similar approach.)
- Talk through and model the investigation planning process – make predictions, choose equipment, plan the method, work out what results to collect and how to record them – and how to use results to draw a conclusion.
- Give the learners planning time in pairs or small groups in order for them to decide what to do and how to do it.
- Use photocopiable pages 19–22 as a guide. You could work through these systematically as a class if you are unsure of the confidence levels of the learners in planning and carrying out practical activities at the start of the school year. Give photocopiable pages 19 and 20 to the learners who need support and photocopiable pages 21 and 22 to all the other learners.
- Approve their plans before allowing them to carry out the activity. Prompt them with questions if their ideas lack clarity. They need to be able to perform this activity with a measure of success.

Plenary

- Invite pairs or small groups of learners to share their findings with the whole class.
- Compare and discuss different methods and ways of recording results.
- Agree the qualities of good reflective materials – they are shiny, reflect a torch beam and allow the learners to see themselves in them.
- Discuss which might be the best reflective material from those available and / or tested.

Success criteria

Ask the learners:

- Which materials did you use?
- How did you test them?
- Which material was most reflective?
- What kinds of materials are not good at reflecting light?

Ideas for differentiation

Support: Give these learners a set of three pre-selected materials to test and work with them in a small group to carry out the practical activity.

Extension: Ask these learners to test all the best reflective materials identified by the class and to demonstrate their findings.

Name: _____

Which materials best reflect light? 1

Your teacher will give you three different materials or objects.

Predict

1. Write them in the order that you think will be the most reflective to the least reflective.

    most reflective

    a) _____

    b) _____

    c) _____

    least reflective

Equipment

2. List the equipment you will need:

    - _____
    - _____
    - _____
    - _____

Method (what will you do?)

3. Draw or write to show how you will do your experiment.

Name: _____

Which materials best reflect light? 2

Results (what happened?)

4. Decide how you will record what you find out.
   Use this table to help if you need to.

Material	

Conclusion (what you found out)

5. Which material was the most reflective? _____

6. Write a reason why.

   _____

   _____

   _____

Name: _____

Which materials best reflect light? 1

Predict

1. Choose at least five materials and put them in the order you think will be the most reflective to the least reflective:

   most reflective

   a) _____

   b) _____

   c) _____

   d) _____

   e) _____

   least reflective

Method (what will you do?)

2. Write or draw how you will carry out the experiment.

Cambridge Primary: Ready to Go Lessons for Science Stage 5 © Hodder & Stoughton Ltd 2013

Name: _____

Which materials best reflect light? 2

Equipment

3. List the equipment you will need:

   _____   _____   _____

   _____   _____   _____

   _____   _____   _____

Results (what happened?)

4. How will you record what happened? (Use the back of this page if you run out of space.)

```
┌─────────────────────────────────────────────────────────────┐
│                                                             │
│                                                             │
│                                                             │
│                                                             │
│                                                             │
└─────────────────────────────────────────────────────────────┘
```

Conclusion (what you found out)

5. Which materials were the most reflective?

   _____

   _____

   Why? _____

   _____

Unit 1A: 5.1 The way we see things

Making periscopes

Learning objectives

- Know that scientists have combined evidence with creative thinking to suggest new ideas and explanations for phenomena. (5Ep1)
- Collect sufficient evidence to test an idea. (5Ep5)
- Know that beams / rays of light can be reflected by surfaces including mirrors, and when reflected light enters our eyes we see the object. (5Pl7)

Resources

Plane mirrors; internet access; scissors; sticky tape; card; empty (clean) milk or orange-juice cartons; protractors or angle measurers; photocopiable page 24.

Starter

- Give each pair of learners a mirror. Ask them to hold the mirror in front of them and have their partner stand behind them. Where do they need to hold the mirror in order to see their partner?
- Repeat this activity, asking the partner to position themselves to one side, to the other side, further back, or closer. Each time the person with the mirror has to move it until they can see their partner in it.
- Discuss their findings each time.
- Challenge them to use their mirror to enable them to see around a corner.

Main activities

- Explain that sometimes it is useful to be able to see around corners. Discuss any suggestions that the learners might have. They might include some fun elements here, for example looking at something whilst they remain unseen!
- Ask if any of them know a piece of equipment that has been designed to do just that (a periscope). Show a picture of a periscope or film footage of a periscope being used, for example on a submarine. Discuss what events might have prompted the need for such a piece of equipment. If you have a ready-made periscope available, allow some of the learners to come forward and use it. Ask them to describe the experience to the rest of the class.
- Discuss how they might make a periscope – what equipment they might need and how they might construct it.
- Demonstrate how to make a periscope – give out and follow the instructions on photocopiable page 24. Check that the learners are capable of using protractors and / or angle measurers. If not, spend some time going over this with them. Ask some of the learners to help you in constructing a periscope and using it when it is made.
- Allow the learners to work in pairs to help each other perform the task of making their own periscopes.

Plenary

- Allow the learners to use the periscopes to identify objects around the room that are behind screens (these can be made simply from tall piles of books).

Success criteria

Ask the learners:
- How does the periscope work?
- What can it be used for?
- What was the most difficult part in making the periscope?
- What might you use a periscope for?

Ideas for differentiation

Support: Arrange for these learners to work in pairs with learners who do not need support.

Extension: Challenge these learners to set up objects around the room to be viewed by their classmates using the periscopes that they have made.

Making a periscope

You will need:

Scissors, sticky tape, 2 plane mirrors, card, 2 empty boxes (milk cartons are ideal), a protractor or angle measurer.

What to do

- Cut the tops from both the cartons.
- Place a piece of card at the bottom of both the cartons at a 45 degree angle and fix it with sticky tape.
- Put a mirror on each of the pieces of card and secure it with more sticky tape.
- Cut out a small square hole directly in front of the mirror in each of the cartons.

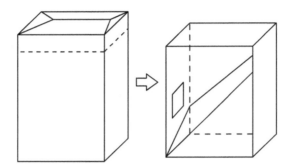

- Place the two cartons on top of each other – open ends together. Tape them together.

IMPORTANT!

Make sure that the mirrors are placed at a 45 degree angle (otherwise the periscope won't work).

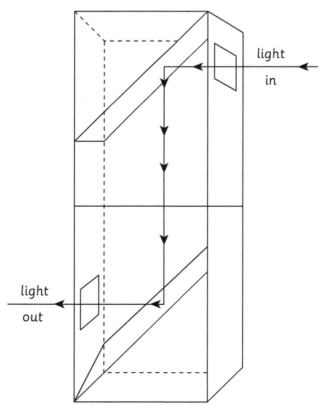

Unit 1A: 5.1 The way we see things

Light beams

Learning objectives

- Make predictions of what will happen based on scientific knowledge and understanding, and suggest and communicate how to test these. (5Ep3)
- Explore why a beam of light changes direction when it is reflected from a surface. (5Pl8)

Resources

Flipchart and markers; plane mirrors; small torches and spare cells (batteries) or laser pens (if available – laser pens show light travelling in coloured lines, but should not be shone in the eyes; decide whether to allow the learners to use them or just the teacher for demonstration purposes); paper; rulers; coloured pencils; sticky tape or sticky tack; photocopiable page 26.

Starter

- Ask the learners to discuss with talk partners what things give off beams or rays of light.
- Discuss and listen to the learners' responses. (These could include such things as the Sun, the Moon, torches, car headlights, TVs, computer monitors.)
- List all suggestions made.
- Explain that all light sources send out light, some as beams or rays (as discussed), but others send out light in every direction from the source, for example a candle flame, the Sun, an electric light in a room. Some natural things give off light, for example sunbeams. However, moonbeams are a result of sunbeams reflecting off the Moon. The Moon itself is not a light source; it reflects the Sun's rays. The Moon acts like a huge mirror in space. Light rays from the Sun bounce off the Moon and travel to us on Earth.

Main activities

- Explain to the learners that in this lesson they will find out about how light is reflected by a plane mirror.
- Demonstrate what the learners have to do, following the instructions given on photocopiable page 26. Ask some learners to assist you and then give them time in mixed-ability groups to carry out the activity. Ensure that they predict what they think might happen before they carry out the activity.

Plenary

- Invite the learners to share their results with the rest of the class.
- Discuss their findings and address any misunderstandings. Explain that the reflected angle is always the same, that is, equal to the angle at which the beam hits the mirror. (The learners do not need to know the term 'angle of incidence' here.)

Success criteria

Ask the learners:

- What did you see?
- What happened when you changed the position of the torch?
- Did you see a pattern in your results?
- What does this tell us about how light travels even after it has been reflected? (It still travels in a straight line.)

Ideas for differentiation

Support: Allow these learners to work in mixed-ability groups, or work in a small group with them.

Extension: Challenge these learners to find a way of using a mirror to reflect a torch beam onto a specific place, for example the classroom ceiling or a poster on the wall. Ask them to demonstrate this to the rest of the class.

Name: _____

Beams of light

You will need:

A small light source (torch or laser pen), a plane mirror, sticky tack, paper, coloured pencils, ruler.

What to do

- Set up the equipment as shown in the diagram below, making sure that your mirror is standing on a piece of paper.

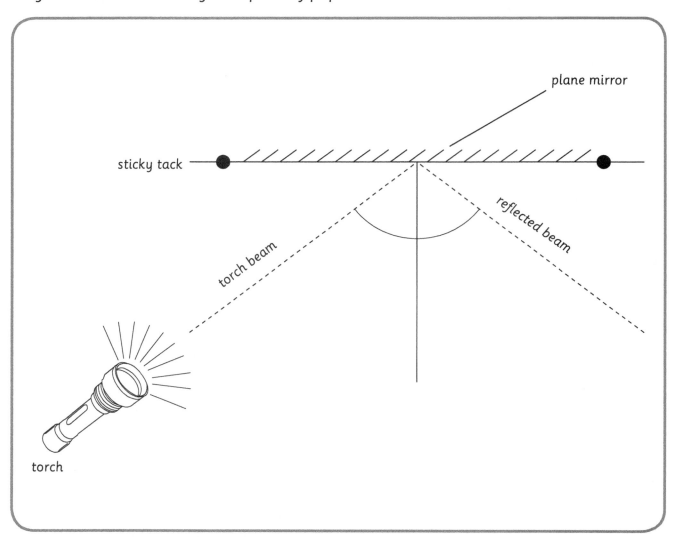

- Shine the torch or laser pen at the mirror and look at the reflected angle of light. Draw both lines on the paper using a coloured pencil and a ruler.

- Move the torch and repeat the activity using a different coloured pencil.

- Do this again at least three times – use a different colour each time.

What do you notice about the reflected beam of light each time?

_____

Unit 1A: 5.1 The way we see things

How many images?

Learning objectives

- Present results in bar charts and line graphs. (5Eo4)
- Recognise and make predictions from patterns in data and suggest explanations using scientific knowledge and understanding. (5Eo7)
- Know that beams / rays of light can be reflected by surfaces including mirrors, and when reflected light enters our eyes we see the object. (5Pl7)

Resources

Plane mirrors; a small figure or toy; sticky tape and sticky tack; protractors or angle measurers; photocopiable page 28.

Starter

- Show the learners two mirrors joined with sticky tape to look like the pages of a book.
- Ask them with talk partners to predict what they would see if they looked into that mirror.
- Allow them to tape two mirrors together to try it for themselves. Were they correct in their predictions?

Main activities

- Give out photocopiable page 28. Discuss and go through the activity with the learners. Explain that they have to count the number of images they can see in the mirror each time when the mirror is opened at different angles.
- Carry out the activity in pairs or small groups. Ensure that the learners are confident in using a protractor or angle measurer. If not, demonstrate and give them time to practise using one.
- Discuss the results once the activity is completed, with reference to the completed results tables on photocopiable page 28.
- Demonstrate how to turn these results into a simple graph, as below. Plot the angle on the x-axis and the number of images on the y-axis. Discuss the importance of a title, appropriate scale and labelling of axes. Plot points as either a cross or a point with a circle around them.

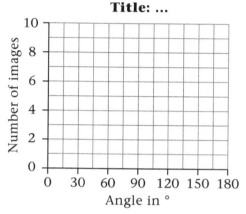

- Ask the learners to draw a similar graph of their own results on photocopiable page 28.

Plenary

- Discuss and compare individual graphs. Ask similar questions of different individuals, pairs or small groups to help them to interpret their graphs, using the success criteria questions to assist you.

Success criteria

Ask the learners:
- Which angle gave the fewest images?
- What is the shape of the resulting graph?
- What does this tell us?
- Is there a formula we could apply for the number of images seen for any angle?

Ideas for differentiation

Support: Allow these learners to work in mixed-ability groups or pairs, or work with them in a small group. Help them to use the angle measurer or protractor. Advise them of the correct scale to choose for their graph. Check graphs as each point is plotted.

Extension: Challenge these learners to work out a formula that gives the number of images for any angle.

Name: _____

How many images?

You will need:
Two plane mirrors taped together, sticky tack, a protractor or angle measurer, a small toy figure.

What to do
- Place the mirrors at an angle like an open book.
- Place the toy in front of the mirrors.
- Measure the angle of the mirrors.
- How many images can you see reflected in the mirrors?
- Repeat the activity for the angles in the table below and record the number of images you see.

Results

Angle	Number of images
180°	
90°	
60°	
45°	
30°	

Draw a graph showing your own results below. Remember to include a title and to label the axes.

Title: Graph to show _____

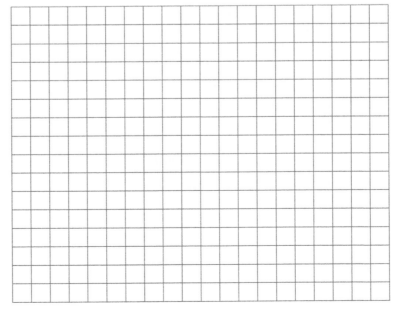

Unit 1A: 5.1 The way we see things

Comparing reflections

Learning objectives

- Make relevant observations. (5Eo1)
- Explore why a beam of light changes direction when it is reflected from a surface. (5PI8)

Resources

Plane mirrors; aluminium foil; a collection of items with shiny surfaces, e.g. metal spoons, polished wood; bowls of water; photocopiable pages 30 and 31.

Starter

- Give the learners in small groups time to explore how many of the collection of objects enable them to clearly see a reflection in them.
- Discuss and share findings with the whole class. Compare any disagreements and come to an agreement before continuing. Ask: *What type of surface makes a good clear reflection?*

Main activities

- Explain that in this lesson the learners will find out which surfaces make better reflective surfaces than others. Discuss what the word 'reflective' means. Ask the learners to identify reflective surfaces around the classroom. Ask them to think about how they might describe a reflective surface, which will help them to concentrate on observation during the activity/ies.
- Give out and go through the activities on photocopiable pages 30 and 31. Answer any questions that arise and ensure that the learners are confident in what they have to do before they begin the practical activity/ies. Organise the learners into mixed-ability groups to carry out the practical activities. Alternatively, organise the groups so that the learners who need support have an adult helper working with them.

- Discuss the learners' findings from the experiments. Ensure that they all understand that we are able to see things only because light is reflected from surfaces into our eyes. It is important to remind the learners that the surfaces of all objects reflect light – it may be a little or it may be a lot. This is a difficult concept for many learners and needs to be re-emphasised regularly throughout this unit of work. Many learners will find it difficult to understand that we see things because of the light given off by objects.

Plenary

- Discuss the fact that smooth polished surfaces reflect light well. Introduce the idea that non-shiny surfaces absorb light and do not reflect it.
- Compare the differences seen in reflections made by the crumpled foil and shaken water. Explain that these reflections are less clear because the light from their rough surfaces gets scattered in all directions.

Success criteria

Ask the learners:

- Name something that you have used in today's lesson that reflected light well.
- Describe the difference between the reflections in flat and in crumpled foil.
- What was different about the reflection seen in the shaken water?
- How does the surface affect how clear the reflection is?

Ideas for differentiation

Support: Supervise these learners working in a small group or allow them to work in mixed-ability groups. Closely supervise all the activities involving the bowl of water.

Extension: Challenge these learners to look at reflections in the back and front of a metal spoon. What do they notice? Ask them to report back to the rest of the class on what they see.

Name: _____

Comparing reflections 1

Looking in a mirror

You will need:
A plane mirror, a flat piece of aluminium foil.

What to do
- Look at yourself in the mirror.
- Draw what you see in the table below.
- Look at yourself in the flat piece of foil.
- Draw what you see in the table below.
- Compare and talk about what you have seen with your talk partner.

Mirror	Foil

- How would you describe the differences between the two pictures?

_____

_____

Name: _____

Comparing reflections 2

You will need:
A plane mirror, a flat piece of aluminium foil, a small bowl of water.

Crumpled foil

What to do
- Squeeze the foil into a tight ball, then make it flat again.
- Draw what you see now.

Reflections in water

What to do
- Place the bowl of water on the table in front of you.
- Wait until the water settles.
- Look into it.
- Draw what you see.
- Repeat this activity, but shake the bowl before drawing the second picture.

Crumpled foil

Still water	Rippling water

Cambridge Primary: Ready to Go Lessons for Science Stage 5 © Hodder & Stoughton Ltd 2013

Unit 1A: 5.1 The way we see things

Optical illusions

Learning objectives

- Know that scientists have combined evidence with creative thinking to suggest new ideas and explanations for phenomena. (5Ep1)
- Know that we see light sources because light from the source enters our eyes. (5Pl6)

Resources

Internet access; photocopiable pages 33 and 34.

Starter

- Type 'optical illusions for kids' into your search engine.
- Select a sample of activities for the learners to try out. Be selective and familiarise yourself with some of the activities before the lesson.
- Discuss why the learners think they are tricked into seeing something that is not really there. This is a lot of fun – and prompts many interesting questions from the learners.

Main activities

- What are illusions? Explain that optical illusions trick us into seeing something different from the way it actually is, so that what we see does not match the real world. As you will have seen, in some illusions one person sees something in a picture, while in the same picture someone else sees something entirely different. There will probably be some lively discussion at this stage about who is right.
- Use this point to bring out the fact that research scientists must be sure that the results of their work are not illusions. They need to accurately report what actually is, not just their general impression of what is. Scientists repeat experiments many times, or in different laboratories, to obtain valid and reliable results. Science is only good Science when anyone can repeat the experiment and get the same results anywhere in the world.
- Please note, some optical illusions may cause dizziness or possibly epileptic seizures. This is because the brain cannot handle the conflicting information from our two eyes. If any learners start to feel unwell when doing these activities, they should immediately cover one eye with their hand and then stop. They should not close their eyes because that can make the attack worse.
- Give out photocopiable pages 33 and 34, which provide some paper-based optical illusions.

Plenary

- Take time to discuss in detail what different people perceive. Old or young lady? The old lady is looking towards the left with her nose in the centre. The young lady is in profile, looking away into the distance. Your eyes see black spots inside the circles on the edge of your peripheral vision – this is a different distance for everyone, so there will be varied responses.
- Explain that this is because of the way in which the different parts of our brain (right and left sides) process information. Discuss how different types of learners have different areas of their brains that appear more developed, which gives them skill in particular areas, for example scientific, logical thinking or creative thinking.

Success criteria

Ask the learners:

- Which illusion did you see easily?
- Which illusion puzzled you?
- Did you agree with your friends every time?

Ideas for differentiation

Support: Allow more time for these learners to assimilate what they might see.

Extension: Challenge these learners to find further examples of optical illusions from the internet.

Name: _____

Optical illusions 1

1. Look at these pictures. What can you see?

   a) Old or young lady?
   _____

   b) Is the book facing towards you or away?
   _____

   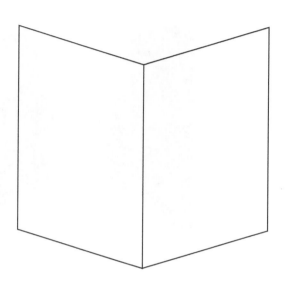

   c) How many black circles can you see?
   _____

Re-drawn after images on www.brainbashers.com.

Name: _____

Optical illusions 2

1. Look at these pictures. What can you see?

   a) A vase or two people?
   _____

   b) A duck or rabbit?
   _____

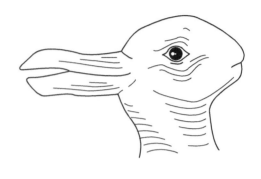

   c) Are these horizontal lines parallel?
   _____

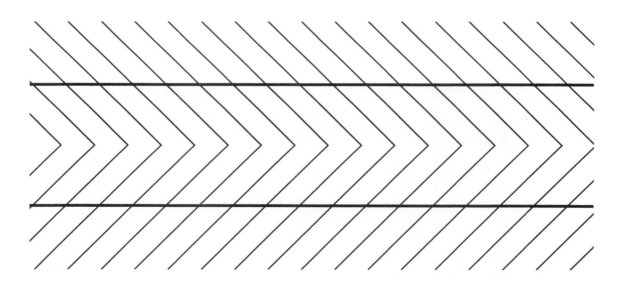

Re-drawn after images on www.brainbashers.com.

Unit 1A: 5.1 The way we see things

Unit assessment

Questions to ask

- How do we see?
- What happens in our eyes to help us to see?
- Name three reflective surfaces or materials.
- What happens when light is reflected by a plane mirror?
- Why can we see our reflection in still water?
- How does crumpling up a piece of aluminium foil affect the light that is reflected from its surface?

Summative assessment activities

Observe the learners while they participate in these activities. You will quickly be able to identify those who appear to be confident and those who may need additional support.

Light sources

This activity will assess the learners' knowledge of different light sources, and their classification as natural or artificial.

You will need:

A selection of different light sources (as used in the lessons) or pictures of a range of different light sources, e.g. torches, candles, oil lamps, pictures of the Sun, lightning; access to computers; poster paper; art materials, e.g. markers or paint and painting equipment or an ICT design package that the learners are confident in using.

What to do

- Ask the learners to group or classify the light sources and pictures of light sources as natural or artificial.
- Ask them to include any more that they can think of.
- Now challenge them to produce a poster to show their findings.

Light rays

This activity assesses the learners' understanding of light reflection by different surfaces.

You will need:

A plane mirror; ruler; pencil; paper; a selection of opaque, transparent and translucent objects or materials.

What to do

- Ask the learners to draw diagrams to show how light rays are reflected by an opaque, a transparent and a translucent surface. Remind them to draw arrows on the light rays.

OR

- As teacher, present pre-drawn diagrams and ask the learners to identify which is which, that is, opaque, transparent or translucent (or any combination of these three).

Written assessment

Distribute photocopiable page 36. The learners should work independently, or with the usual adult support they receive in class.

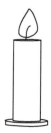

35

Name: _____

The structure of the eye

1. Use this diagram to help you complete the sentences about the eye.

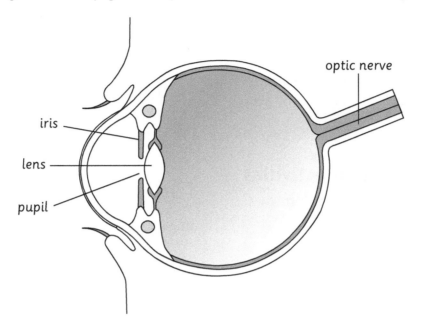

a) The black circle in the centre of the eye is called the _____.

b) The eye is connected to the brain by the _____ _____.

c) The part of the eye that helps us to focus is the _____.

d) The coloured circle in the centre of the eye is called the _____.

2. Answer these questions.

a) What does the pupil do?
_____

b) What happens when the eye focuses on something?
_____
_____

c) Why are two eyes better than one?
_____
_____
_____

Unit 1B: 5.2 Evaporation and condensation

What is evaporation?

Learning objectives

- Make relevant observations. (5Eo1)
- Measure volume, temperature, time, length and force. (5Eo2)
- Know that evaporation occurs when a liquid turns into a gas. (5Cs1)

Resources

A bottle of perfume or air freshener (or something strongly scented – ensure that you check for allergies, especially if using air fresheners); a selection of pictures (from books / posters / the internet) with examples of evaporation in everyday life; heat source, e.g. a candle, a nightlight or a spirit burner; container of water, e.g. a cup of boiling water with steam coming off it; measuring cylinder; photocopiable page 38; a large space.

Starter

- Ask the learners to close their eyes, then leave a bottle of perfume, fragrance or something that is strongly scented open near to them for a few minutes. Ask them to indicate when they think they know why you have asked them to close their eyes.
- With talk partners, ask the learners to discuss why they are able to detect the fragrance (it evaporates into the air).
- Talk about other things that they like the smell of, or even unpleasant smells that they sometimes detect.
- Either as a class, or in pairs with talk partners, ask the learners to identify what is happening in each of the pictures. These can be displayed for the whole class to see, or distributed as smaller pictures around the class to invite comments and participation from more learners.
- Ask the learners to draw a picture of how the scent reached them. Discuss their responses and correct any misconceptions.

Main activities

- Explain that the work in this unit builds on what the learners have already learnt in Stage 4 Solids, liquids and gases. This means that some of the work will be revision. You need to find out how much they have remembered before moving on to the processes of evaporation and condensation.
- Demonstrate water boiling and the evaporation that takes place, using the resources available to you. Ask the learners to measure the initial volume using a measuring cylinder. For safety reasons, you will need to measure the volume of the residual water yourself (or have an adult helper do it). Discuss the difference.
- During the demonstration, discuss what is happening in terms of the states of matter involved – the liquid (perfume or water) turns into a gas by evaporation. Revise the qualities of:
  - liquids – they take the shape of the container they fill, cannot be cut, flow easily
  - gases – they can be squashed and fill any space they occupy.
- Give out photocopiable page 38 and explain what the learners have to do to complete it.

Plenary

- Emphasise that liquids evaporate and become gases, and when water evaporates we call it water vapour.
- Role-play being particles in a liquid, being heated up and evaporating.

Success criteria

Ask the learners:

- What is evaporation?
- What do you notice when steam is produced?
- How are the particles arranged in a liquid compared to in a gas?
- What is the difference in the volume of water before and after evaporation?

Ideas for differentiation

Support: Work with these learners in a small group to complete photocopiable page 38.

Extension: Ask these learners: *Can you think of any everyday examples of water evaporating?*

Name: _____

What is evaporation?

Use these words to help you complete this page (you will not need them all).

| gas | heat | liquid | particles | solid | steam | water vapour |

1. Complete the sentence below.

   Evaporation happens when a _____ turns into a _____.

2. Draw the particles in a liquid and in a gas in the table below.

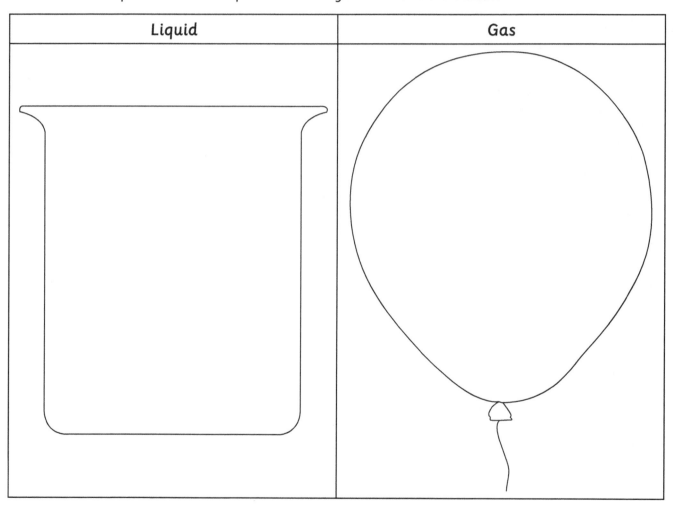

3. What is needed to make a liquid turn into a gas? _____

4. Liquids evaporate to become gases. What does water evaporate to become?
   _____ _____

5. What is another word for water vapour? _____

Unit 1B: 5.2 Evaporation and condensation

The water cycle

Learning objectives

- Know that evaporation occurs when a liquid turns into a gas. (5Cs1)
- Know that condensation occurs when a gas turns into a liquid and that it is the reverse of evaporation. (5Cs2)
- Measure volume, temperature, time, length and force. (5Eo2)

Resources

Beakers or jars; water; measuring cylinders; internet access; poster or books containing a diagram of the water cycle; heat source; tile or mirror; poster paper; collage or art materials; scissors; glue; pre-printed vocabulary labels; photocopiable page 40.

Starter

- Set up a container of water in a warm place. Measure the volume of water at the beginning and at regular intervals throughout the day. Record the results. Discuss after some of it has evaporated. Ask the learners to think about how you might recover the lost water. Ask them to predict how quickly the water might all evaporate.
- Show a film clip of the water cycle from the internet (if available). There are several examples on YouTube – select the level of description that you feel is appropriate for your learners.

Main activities

- Recap on evaporation from the previous lesson. Ask the learners: *What is evaporation?* Demonstrate rapid evaporation again by boiling water. Measure the starting volume of water used. Show the learners that water vapour can be turned back into water droplets by being cooled down. Hold a mirror or tile in the water vapour and observe the droplets forming.
- Introduce the term 'condensation' and describe what it is. Measure the volume of condensed water collected. Discuss findings. Ask the learners to give any examples from everyday life when they have seen evidence of evaporation or condensation, for example puddles drying up, clothes drying, condensation in the bathroom on tiles or windows.
- Introduce the relevant scientific vocabulary associated with the water cycle – evaporation, condensation, precipitation, run-off.
- Talk through the water cycle using these terms and discuss what happens at each stage.
- Give out photocopiable page 40 for the learners to complete.

Plenary

- Invite the learners to share their completed work with the rest of the class. Praise them for the correct use of specific scientific vocabulary. Emphasise the importance of being able to spell these words correctly.
- Discuss the inverse relationship between evaporation and condensation, which is that condensation is the reverse of evaporation.

Success criteria

Ask the learners:

- What is evaporation?
- How does condensation form?
- What is the scientific name for rain?
- Are the processes of evaporation and condensation related in any way?

Ideas for differentiation

Support: Make a group collage or poster of the water cycle with these learners. Ask them to label the stages using pre-printed labels. Display and show the completed work to the rest of the class.

Extension: Ask these learners to research useful examples of evaporation around the world, for example salt production by evaporation of salt pools.

The water cycle

Name: _____

Label and colour this diagram of the water cycle.

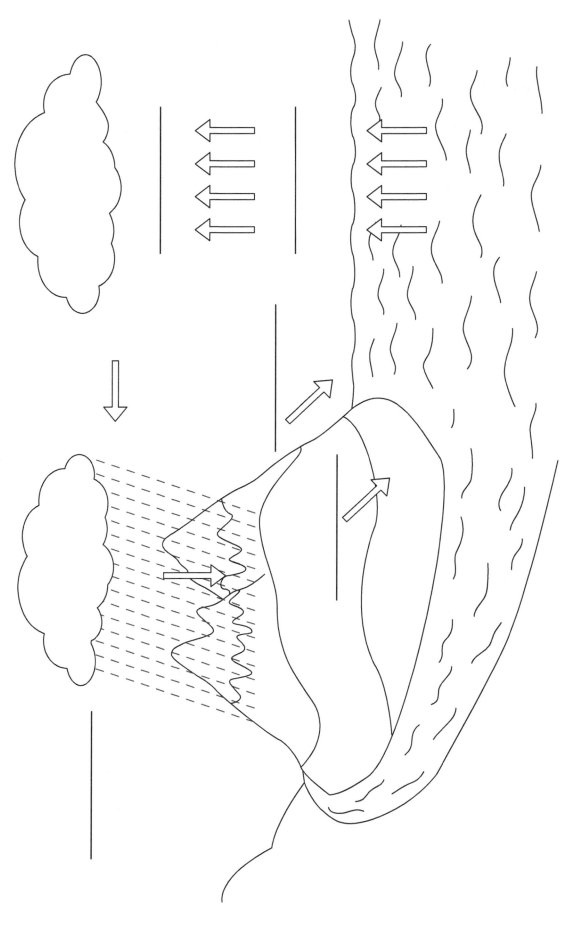

Unit 1B: 5.2 Evaporation and condensation

What is condensation?

Learning objectives

- Make relevant observations. (5Eo1)
- Know that condensation occurs when a gas turns into a liquid and that it is the reverse of evaporation. (5Cs2)
- Know that air contains water vapour and when this meets a cold surface it may condense. (5Cs3)

Resources

Heat source; container of water; tile or mirror; poster of the water cycle; cling film / plastic wrap; beakers or containers for hot water; ice cubes; measuring cylinders; photocopiable pages 42, 43, 44 and 45.

Starter

- Ask the learners to explain what happens when water vapour condenses. (It turns back into liquid water.) Demonstrate this again (as in the previous lesson) by boiling some water and allowing the water vapour to condense on a cold surface to remind them.
- Refer to the poster produced by the support group in the previous lesson. Ask the learners with talk partners to identify the stages in the water cycle where evaporation and condensation occur. Discuss their answers, making relevant comments and asking appropriate questions where necessary.
- Re-emphasise that:
  - water vapour is invisible
  - water evaporates from seas, rivers, lakes, animals and plants – usually by the action of the Sun's heat on them
  - water vapour condenses and forms clouds.

Main activities

- Give out photocopiable pages 42 and 43 to all the learners except those who need extension; give these learners photocopiable pages 44 and 45. Demonstrate how to set up the experiment to be able to observe the process of condensation. Ensure that any hot water the learners use is only hand-hot, that is, no more than 40°C.
- Organise the learners to work in pairs or small groups, depending on the equipment and time available. Allow them to follow the instructions on photocopiable pages 42–45 to set up and carry out the experiment, completing the responses as they work. Circulate around the room, observing, assisting and asking relevant questions as necessary.

Plenary

- Discuss what happened and what the learners observed.
- Discuss why condensation occurs on cold taps in the bathroom or kitchen or on cans of drink taken from the fridge. (Gas touching a cold surface condenses back into a liquid.) Ask the learners to identify other examples and discuss their suggestions.

Success criteria

Ask the learners:

- What did you observe?
- Why did this happen?
- Describe the droplets and what happened to them.
- What happens to water vapour when it hits a cold surface?

Ideas for differentiation

Support: Supervise these learners in a small group, or organise them into mixed-ability groups so that their friends can help them.

Extension: Give these learners photocopiable pages 44 and 45 to work from. Challenge them to measure the original volume of hot water used and the final volume remaining at the end of the experiment. Ask them to share their findings at the end of the lesson. How can they explain any differences in the volume of water?

Name: _____

What is condensation? Experiment

You will need:
Ice cubes, cling film / plastic wrap, a beaker, hot water.

What to do
- Pour some hot water into the beaker.
- Stretch cling film / plastic wrap tightly over the top of the beaker. (Be careful – the water will be hot!)
- Place one or two ice cubes on top of the cling film.
- Observe and record what happens.

Diagram
Draw a labelled diagram showing how you set the experiment up.

Name: _____

Condensation experiment results

1. Touch the cling film gently in different places. Is it warmer or colder in some places than in others?

2. Shade the diagram to show any differences in temperature on the cling film.

3. Where did you see the biggest droplets?

   _____

   _____

4. Where were the smallest droplets?

   _____

   _____

5. Explain why you think this happened.

   _____

   _____

Cambridge Primary: Ready to Go Lessons for Science Stage 5 © Hodder & Stoughton Ltd 2013

Name: _____

What is condensation?
Experiment

You will need:
Ice cubes, cling film / plastic wrap, a beaker, hot water, a measuring cylinder.

What to do
- Pour a measured amount of hot water into the beaker. Record how much water you have used.
- Stretch cling film tightly over the top of the beaker. (Be careful – the water will be hot!)
- Place one or two ice cubes on top of the cling film.
- Observe and record what happens.

Diagram
Draw and label a diagram showing your set-up.

Name: _____

Condensation experiment results

1. Observe closely what happens.

2. Record what you see and describe where you see it. Choose the best way to show this – will you draw, write or do both?

3. When the ice cubes have melted, look at the volume of water in the beaker.

   a) Is it the same as or different from at the start? _____

   b) Explain why you think this is. _____
   _____
   _____

Cambridge Primary: Ready to Go Lessons for Science Stage 5 © Hodder & Stoughton Ltd 2013

Unit 1B: 5.2 Evaporation and condensation

Hot or cold?

Learning objectives

- Measure volume, temperature, time, length and force. (5Eo2)
- Discuss the need for repeated observations and measurements. (5Eo3)
- Know that the boiling point of water is 100°C and the melting point of ice is 0°C. (5Cs4)

Resources

Icy water; cold water; water at room temperature; warm water; hot water; bowls for water; towels; labels describing each type of water being used; thermometers; adhesive tape or sticky tack; ice cubes; photocopiable page 47.

Starter

- Have ready some bowls of water at different temperatures, but not labelled. Invite the learners to come forward and compare the water in any two bowls by placing their hand in the bowl of water for about ten seconds (or longer). Which is hotter? Which is colder? Do not use water hotter than hand-hot (no more than 40°C) – test the temperature with your inner wrist.
- Work through this activity using different learners until you have a set of bowls of water arranged from coldest to hottest. Label the bowls.
- Ask another learner to compare the hottest and coldest bowls of water by placing one hand in each bowl at the same time. Ask them to then put both hands into a bowl of warm water and to describe what they feel. They should say that the hand that has been in the icy water now feels warm, but the hand that has been in the hot water feels cooler.
- Use their response to explain that hot and cold are relative measures – not exact measurements. In this lesson they will find out some exact measurements of how hot or cold some liquids are.

Main activities

- Introduce the word 'temperature' as a measure of how hot or cold something is and explain that it is measured using a thermometer.
- Demonstrate safe and good handling of a thermometer. It needs to be removed from its storage tube and returned to it again after use to protect it. To prevent a thermometer rolling off the table, secure a piece of adhesive tape or sticky tack to the table surface and use this to hold the thermometer in place whilst it is needed.
- Give out thermometers. Describe the scale on the thermometer and introduce the vocabulary 'degrees Celsius' and the notation °C.
- Identify what the room temperature is and discuss any differences in readings and possible reasons why, for example it is hotter by the window in the Sun. Circulate the room and check that the learners are reading the scale correctly.
- Give out photocopiable page 47 and explain to the learners that they have to use this page to record the temperatures they take.

Plenary

- Go through the learners' responses to the questions on photocopiable page 47. Discuss their findings and consider any anomalies.
- Emphasise the need for taking repeat measurements for reliability.

Success criteria

Ask the learners:

- Where is the hottest part of the room?
- What temperature is it?
- What is the temperature of ice?
- Predict the temperature of boiling water.

Ideas for differentiation

Support: Give these learners a thermometer with an easy-to-read scale.

Extension: Challenge these learners to find out about the hottest and coldest places in the world.

Name: _____

Taking temperatures

You will need:

A thermometer, an ice cube, icy water, cold water, warm water, hot water.

What to do

- Use the thermometer to take temperatures.
- Take each temperature three times.
- Complete the table of results.

Object or substance	Temperature in °C		
	1	2	3
Air			
Ice cube			
Icy water			
Cold water			
Warm water			
Hot water			

- The coldest thing was: _____
- The hottest thing was: _____

Unit 1B: 5.2 Evaporation and condensation

Melting and boiling points of water

Learning objectives

- Measure volume, temperature, time, length and force. (5Eo2)
- Begin to evaluate repeated results. (5Eo6)
- Know that the boiling point of water is 100°C and the melting point of ice is 0°C. (5Cs4)

Resources

Thermometers; kettle or heat source for boiling water; ice cubes; temperature sensor (if available); graph paper; photocopiable pages 49, 50 and 51.

Starter

- Recap from the previous lesson what the learners measured the temperature of ice to be.
- Discuss any differences in the figures obtained.
- Ask the learners to discuss with talk partners if there is a better way to take the temperature of ice.
- Discuss their responses.

Main activities

- Show the learners an enlarged version of photocopiable page 49, which shows a set of results for the temperature of water as it boils. Ask them to predict the next few readings.
- Show them the second set of figures starting from 90°C and again ask them to predict the next few readings. Hand out photocopiable page 49 for the learners to complete.
- Now give out photocopiable pages 50 and 51. Explain that the learners need to record the temperatures of melting ice and boiling water. Go through the methods for doing this as outlined on the pages.
- Demonstrate taking the temperature of boiling water yourself (or have an adult helper do it, for safety reasons). Ask some of the learners to record the temperatures as the water boils, under close adult supervision, and record this on photocopiable page 51.
- Demonstrate how to use a temperature sensor if one is available. Make a mixture of ice and water. Use the temperature sensor to record the temperature of the mixture at 10-minute intervals over one to two hours.

Plenary

- Discuss the learners' results and take an average for the melting and boiling temperatures of water.
- Analyse the results produced by the temperature sensor. *What do these figures tell you about the melting temperature of ice, and room temperature?*
- Predict what the results might be if the room were 10°C hotter or colder.
- Explain that the boiling point of pure water (at standard temperature and pressure – but they do not need to know this) is 100°C. They need to remember this fact.
- The melting point of ice (frozen water), when it turns to liquid from ice, is 0°C (again in standard conditions). They also need to remember this fact.
- Discuss what the bar chart on photocopiable page 51 shows – when the water reaches its boiling point, the graph plateaus.

Success criteria

Ask the learners:
- What is 'temperature'?
- Why is it useful to take an average of results?
- What temperature does ice melt at?
- What is the boiling point of water?

Ideas for differentiation

Support: Work in a small group with these learners, or allow them to work in mixed-ability groups.

Extension: Ask these learners to plot a graph using the results presented for discussion in the Main activities. Ask them to use it to explain what happened to the water when it boiled.

Name: _____

Making predictions
Measuring the boiling point of water

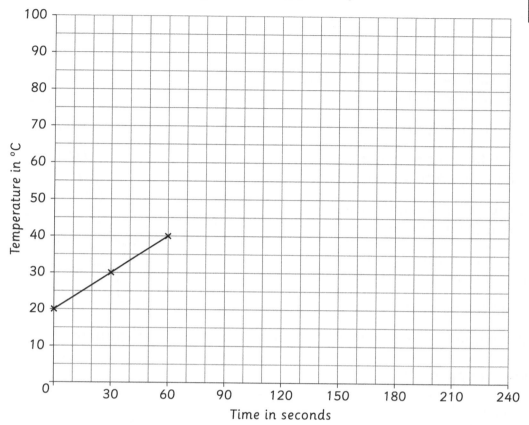

1. Look at the pattern in the graph above.

2. Complete this sentence.

   The temperature goes up _____ °C every _____ seconds.

3. Complete the table below to predict the temperature rise you might expect to see between 70°C and 100°C.

Time in seconds	Temperature in °C
150	70
180	
210	
240	100

Compare your predictions with the actual measurements found in class.

Cambridge Primary: Ready to Go Lessons for Science Stage 5 © Hodder & Stoughton Ltd 2013

Name: _____

Melting point of ice

You will need:
A beaker of ice, a thermometer.

What to do
- Touch the surface of the ice with the bulb of the thermometer.
- Record the temperature of the ice.
- Take three different readings and record the temperature each time.
- Put the thermometer in the ice and leave it for a minute.
- Take the temperature again, with the thermometer still in the ice.
- Record the temperature.
- Take three different readings and record the temperature each time.

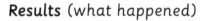

Results (what happened)

Temperature in °C	1	2	3
Surface of ice			
Ice			

Conclusion (what you found out)
What does this show you about the melting temperature of ice (frozen water)?

_____

_____

Name: _____

Boiling point of water

Your teacher will demonstrate water boiling.

1. Use the table below to record the changes in temperature.

Results (what happened)

Time in minutes	Temperature in °C
0	
1	
2	
3	
4	
5	

2. Draw a bar chart of these results below. Remember to include a title and to label the axes.

   Title: Bar chart to show _____

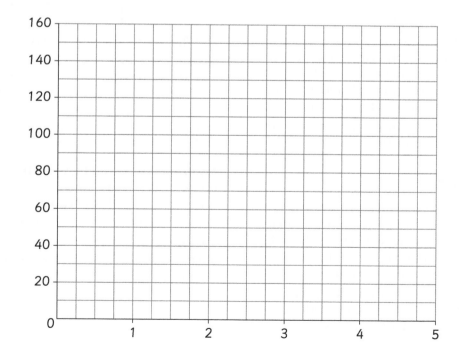

3. What is the boiling point of water from the chart? _____ °C

Cambridge Primary: Ready to Go Lessons for Science Stage 5 © Hodder & Stoughton Ltd 2013

Unit 1B: 5.2 Evaporation and condensation

How long can you keep an ice cube for?

Learning objectives

- Make predictions of what will happen based on scientific knowledge and understanding, and suggest and communicate how to test these. (5Ep3)
- Use knowledge and understanding to plan how to carry out a fair test. (5Ep4)
- Know that the boiling point of water is 100°C and the melting point of ice is 0°C. (5Cs4)

Resources

Ice cubes; salt; string; bowl of water; *Ready to Go Lessons for Science Stage 2* photocopiable page 86 'Ice-cube fishing'; photocopiable pages 53 and 54; selection of materials, e.g. fabric, foil, bubble wrap, paper, plastic; thermometers; timers.

Starter

- Explain that this work builds on work covered in Stage 2 Changing materials and Stage 4 Solids, liquids and gases.
- Play the ice-cube fishing game from *Ready to Go Lessons for Science Stage 2* (photocopiable page 86) as an introductory activity.
- If internet access is available, go to www.bbc.co.uk and search 'bang goes the theory super cool'. This demonstrates how to make water turn into ice immediately with a simple splash or a bang. It is fun to watch.
- Alternatively, watch the film clip yourself and then demonstrate it to the class.

Main activities

- Ask the learners to discuss with talk partners ways in which you could prevent an ice cube from melting for the longest time possible. The only conditions are that you are not allowed to use a fridge or a freezer to keep it in.
- After a few minutes, listen to the learners' suggestions and reasons behind their ideas. Discuss possibilities and ask questions to prompt their thinking.
- Ask the learners in pairs or small groups to plan and design an investigation to find out 'How long can you keep an ice cube for?'
- Give out photocopiable pages 53 and 54, which are planning pages. Move around the room as the learners begin to complete them, offering support and advice as they make suggestions.
- Check and approve the learners' plans before allowing them to carry out their investigation. Offer alternatives to materials they might suggest, if necessary.

Plenary

- Ask the learners, in turn, to share their results with the rest of the class.
- Discuss if the tests were fair – if not, how could they have been improved?
- Decide on the best and worst methods for preventing ice cubes melting.

Success criteria

Ask the learners:
- What did you do?
- How did you make it a fair test?
- For how long did you keep the ice cube before it melted?
- Why did the winning method work?
- What was the longest time it took for an ice cube to melt?

Ideas for differentiation

Support: Work with these learners in a small group or allow them to work in mixed-ability groups. Support them in preparing their plans – either individually, in pairs or as a group.

Extension: Challenge these learners to devise a different method of keeping an ice cube for longer.

Name: _____

How long can you keep an ice cube for? 1

Prediction

1. I think that I will be able to stop an ice cube from melting for

   _____ seconds / minutes.

Method (what will you do?)

2. Write or draw to show what you will need and how you will set it up.

Name: _____

How long can you keep an ice cube for? 2

Results (how will you record what happens?)

3. Write or draw your results below.

Conclusion (what you found out)

4. I kept an ice cube for _____ seconds / minutes.

5. This worked because _____
_____
_____.

Unit 1B: 5.2 Evaporation and condensation

Making solutions

Learning objectives

- Identify factors that need to be taken into account in different contexts. (5Ep6)
- Make relevant observations. (5Eo1)
- Know that when a liquid evaporates from a solution the solid is left behind. (5Cs5)

Resources

Water; fizzy lemonade; milk; white liquid paint; sugar; salt; flour; white powder paint; magnifying glasses; beakers; stirrers; spatulas or spoons; photocopiable page 56.

Starter

- Show the learners the selection of white and colourless solids and liquids available. Explain that they are all familiar everyday liquids and solids, but that they are not all safe to taste. In other words, they are not all foods or drinks.
- Ask the learners with talk partners to group and identify them – without tasting. Prompt them to use the magnifying glasses to help them make more careful observations of each solid or liquid.
- Discuss their groupings and their rationale for the groupings. These reasons could include colour, solids, liquids, food, drink, or non-foods.
- Revise the properties of the solids presented. (Hard / powdery.) Remind them that powders are small grains or particles of solid.
- Revise the properties of liquids. (Can be poured, take the shape of the container, runny, can form drips or drops.)

Main activities

- Explain that in this lesson the learners will make solutions. Ask them what a solution is. (A solid dissolved in a liquid.) At this stage, the liquid is usually water.
- Ask them to suggest solutions that could be made from the solids and liquids available. Allow them to make their suggested solutions. Discuss and compare the results.
- Give out photocopiable page 56. Explain to the learners that they need to follow the instructions and make the solutions.

Plenary

- Discuss each solid and liquid used in turn. Introduce the terms 'solute' (the thing that dissolves) and 'solvent' (the liquid that the solute dissolves in). Compare how effective the solids and liquids used are as a solute or solvent in a solution.
- *Which substances were soluble?*
- *Were there any substances that did not dissolve?*
- Discuss what will happen to each of the solutions made. (The water will evaporate over time.)
- Keep the solutions to look at in the next lesson.

Success criteria

Ask the learners:

- Which factors affect making a solution?
- Which liquid was the best solvent?
- What happens when a liquid evaporates from a solution?
- Name some coloured solids that dissolve in water to make solutions, for example instant coffee, powder paint.

Ideas for differentiation

Support: Limit the number of solutions for these learners to make. Ask them to prepare fewer solutions than the other groups. Use water as the solvent each time for all the solutions you ask them to prepare. Supervise them as they work through the instructions on photocopiable page 56.

Extension: Ask these learners to find examples from everyday life where making solutions is involved, for example in cooking, dishwasher tablets, some medicines (effervescent cures).

Name: _____

Making solutions

1. Complete the word equation below:

   _____ + _____ = a solution

2. Choose from the solids and liquids available and try to make different solutions.

   How will you make this a fair test?

   a) The things I will keep the same each time are (include amounts):

   - _____
   - _____
   - _____
   - _____
   - _____

   b) The **one** thing I shall change each time is _____.

Results

Solid	Liquid	Solution or not? (✓ or ✗)

3. Choose **one** solution and leave it in a warm place. What will happen to it?

   _____

   _____

Unit 1B: 5.2 Evaporation and condensation

Soluble or insoluble?

Learning objectives

- Make predictions of what will happen based on scientific knowledge and understanding, and suggest and communicate how to test these. (5Ep3)
- Decide whether results support predictions. (5Eo5)
- Know that when a liquid evaporates from a solution the solid is left behind. (5Cs5)

Resources

Solutions prepared by the learners in the previous lesson; distilled water, beakers, stirrers, measuring spoons or spatulas; salt; sugar; sand; flour; instant coffee; measuring cylinders; scientific balance; photocopiable page 58.

Starter

- Think back to the solutions made in the previous lesson. Look at them again now. What has happened? (In some of them, the water or liquid will have evaporated to leave the solid behind.) Ask the learners if this is what they thought would happen. Can they explain why this has happened?
- Discuss what has happened to the water or liquid: *Where has it gone?* (It has evaporated into the air; note it has not disappeared.) Remind the learners that this happens because of evaporation.

Main activities

- Recap the vocabulary 'soluble' (dissolves in water or a solvent) and 'insoluble' (does not dissolve).
- Explain that when a solid dissolves in water, it does not disappear; it is still there, and the resulting solution is the solid mixed with water. It might dissolve and produce a colourless solution – but it is still in the water. Remind the learners that it is not safe to taste liquids in Science lessons, unless they are told that the liquids are safe to taste.
- Explain that in this lesson they will make and separate some solutions. Give out photocopiable page 58 and go through the instructions with them. Check their predictions before allowing them to make solutions.
- Discuss making this a fair test – the learners will need to use equal amounts of water, equal amounts of substance and the same number of stirs (if they suggest it) each time. Ask them to think about the best equipment to use – beakers, measuring cylinders, spoons, spatulas, weight in grams of substance.
- Allow the learners to work in pairs or small groups, depending on the amount of equipment available. Provide them with choice from a limited range of suitable equipment.

Plenary

- Compare and share results and findings. Discuss which substances are soluble and which are insoluble.
- Consider which substances will remain after evaporation.

Success criteria

Ask the learners:

- How did you make this a fair test?
- Which substances do you predict will remain after evaporation?
- What remains after the liquid has evaporated?
- Did your results support your prediction/s?

Ideas for differentiation

Support: Work in a small group with these learners or arrange for them to work in mixed-ability groups.

Extension: Ask these learners to rank the substances in order of solubility. How can they compare them?

Name: _____

Soluble or insoluble?

1. Complete the table to show your predictions. Show your teacher.

2. Complete the final column **after** you have made the solutions.

Substance	Prediction – soluble or insoluble?	Was your prediction correct? (yes / no)

Method

3. Remember to carry out a fair test and use water as the solvent.

   a) How much water will you use each time? _____ cm^3

   b) What equipment will you use to measure the water in? _____

   c) How much substance will you use each time? _____

   d) Do your results support your predictions?

   _____

   _____

4. Leave the solutions to see whether evaporation occurs.

   Which substances can you get back from a solution by evaporation?

   _____

   _____

Unit 1B: 5.2 Evaporation and condensation

Growing crystals

Learning objectives

- Identify factors that need to be taken into account in different contexts. (5Ep6)
- Make relevant observations. (5Eo1)
- Know that when a liquid evaporates from a solution the solid is left behind. (5Cs5)

Resources

Hand lenses or magnifying glasses; hot water; salt; sugar; washing soda; jewellery (including crystals); stirrers or glass rods; measuring cylinders; scientific balance; spatulas; beakers or glass jars; cotton thread; scissors; pencils; card; internet access or reference books on crystals; photocopiable page 60; strong heat source.

Starter

- Ask the learners to discuss with talk partners: *What is a crystal?* (A crystal is a substance with a regular shape.) Diamonds could be classed as the most beautiful crystals in the world. They have clear, flat faces that sparkle in the light.
- *Do you know of any crystals? Where would you find crystals at home?* (Salt, sugar, washing soda [take care], jewellery.)
- *Do you know the shape of any crystals?*
- Observe some crystals under a hand lens or magnifying glass. Demonstrate how to use the hand lens, if necessary – hold the lens near to the eye and bring the object and the eye closer together. Good examples to use would be salt or sugar. Salt is a very common seasoning for food.

Main activities

- Explain that most solids are made up of lots of crystals. We usually cannot see them because they are too small. Crystals can be many shapes and sizes. Crystals for a particular substance are always the same shape, for example sugar crystals are always cubes.
- Explain that in today's lesson the learners will grow crystals.

- Give out photocopiable page 60 and go through the instructions with the learners. Demonstrate how to set up the experiment, with help from some of the learners.
- Note that to grow the biggest crystal possible (using your demonstration set-up) you must do the following: follow the instructions on photocopiable page 60 and after a day or two, small crystals should start appearing on the thread around the knot. Pick the best-looking crystal (square and big) and carefully knock off the smaller ones with the back of a knife, taking care not to cut the string. Replace the thread with your chosen crystal in the solution. Keep cleaning the thread of unwanted crystals until you have the crystal you require. As the crystal grows, knock off small side crystals growing on the main crystal. Do not add water to top up the solution, as this will dissolve your crystal.

Plenary

- Demonstrate evaporation of water from a saturated salt solution. Use a strong heat source – a Bunsen burner or a saucepan on a cooker. (Rapid evaporation will leave behind small crystals.)
- Observe the small crystals when cool. *What do you think your crystals will look like?*

Success criteria

Ask the learners:
- What is a saturated solution?
- What will happen to the water over time?
- What shape are sugar crystals?
- Why are the (salt) crystals prepared by the teacher small?

Ideas for differentiation

Support: Work with these learners in a small group.

Extension: Ask these learners to find examples of crystals from around the world.

Name: _____

Growing salt crystals

You will need:
Salt, hot water, three beakers or glass jars, a pencil, thread, card.

What to do

- Tie a large knot in the thread about 1 cm away from the end.
- Tie the thread around the pencil.
- Put the pencil across the top of the first beaker so that the thread does not touch the bottom or sides.
- Make a salt solution in the second beaker by adding salt to hot water. Keep adding salt until no more will dissolve. You should see a layer of salt on the bottom of the beaker. This is called a saturated solution.
- Pour the solution carefully into the third beaker, leaving the undissolved salt in the second beaker.
- Wet the thread with some of the saturated solution and pour the solution into the first beaker.
- Place a piece of card on top of the beaker to stop any dust getting in.
- Put the beaker in the refrigerator or a place where the temperature does not change. This will help the crystals to grow bigger.
- Leave it for a few days until you can see the crystals growing.
- Draw a picture of your biggest crystal in the box below.

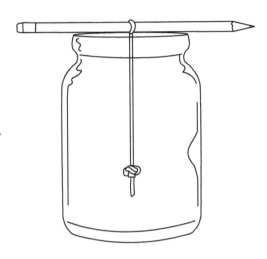

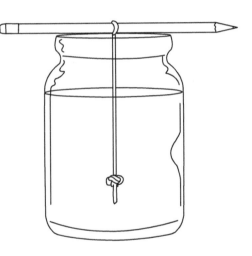

Unit 1B: 5.2 Evaporation and condensation

What's in it?

Learning objectives

- Make predictions of what will happen based on scientific knowledge and understanding, and suggest and communicate how to test these. (5Ep3)
- Decide whether results support predictions. (5Eo5)
- Know that when a liquid evaporates from a solution the solid is left behind. (5Cs5)

Resources

Prepared liquids – (A) saturated salt solution, (B) saturated sugar solution and (C) distilled water; general equipment, e.g. beakers, stirrers, evaporating dishes, heat sources (as required); photocopiable pages 62 and 63.

Starter

- Look at the crystals growing from the last lesson. Ask the learners to observe, compare and describe them. Ask them to draw a crystal if one is clearly visible. Set up a competition for who can grow the biggest crystal. Leave the crystals and look at them over an agreed period of time.
- Show the learners your three prepared liquids. Discuss how they are all the same in appearance. Explain that in Science the learners will work with lots of colourless liquids – but emphasise that it is never a good idea to taste them.
- Ask the learners to discuss with talk partners what the liquids might be and how they could find out. Listen to their responses – but only comment if what they are suggesting is unsafe or dangerous.

Main activities

- Explain that in this lesson the learners' challenge is to be scientific detectives, look for clues and identify each clear liquid.
- Give out photocopiable pages 62 and 63 for the learners to complete. Discuss possible methods, useful equipment and making sure that they carry out a fair test.
- Explain that it is completely their decision which equipment and amounts of substances they use. They will only be able to carry out the experiment once you have approved their plans.
- Allow them to conduct the experiment and report back to the rest of the class during the plenary session.

Plenary

- Ask the learners to identify the three liquids.
- Agree the correct identifications. If they are unable to identify the liquids at this stage, let them know that they can in this instance taste to check (with close adult supervision).
- Ask the learners to describe their methods.
- Decide with the learners if a fair test was carried out each time – make suggestions to improve future testing if not.

Success criteria

Ask the learners:
- What did you do?
- How did you decide what was in each liquid?
- What scientific process has happened to leave a solid behind?
- How did you make your test fair?
- Are all liquids solutions?
- What other solids might dissolve in water to make clear solutions?

Ideas for differentiation

Support: Work with these learners in a small group – guide them through each stage of the planning process. Discuss choices carefully, ensuring that they agree on how to carry out the experiment.

Extension: Give these learners some other solutions and ask them to identify the components. Try some coloured solutions, for example fruit juice, powder paint in water. Ask them to predict and test their predictions.

Name: _____

What's in it? 1

Predictions

What do you think each liquid is?

A _____

B _____

C _____

Equipment (what will you need?)

List the things you will need to use:

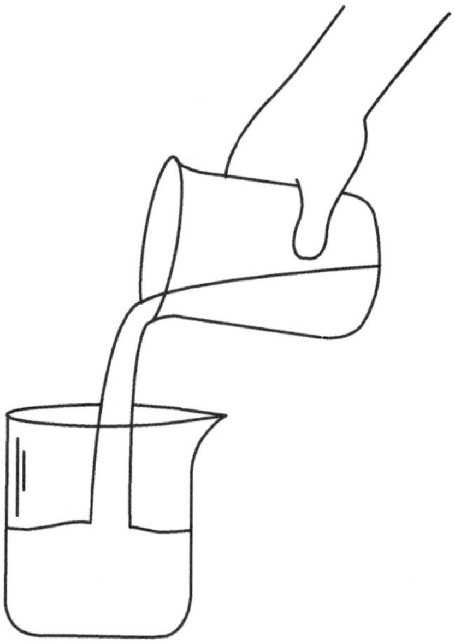

- _____
- _____
- _____
- _____
- _____
- _____
- _____

Method (what will you do?)

1. _____
2. _____
3. _____
4. _____
5. _____

Name: _____

What's in it? 2

Results (what happened?)

Appearance of liquid

A

B

C

Conclusion (what you found out)

1. Which liquid was not a mixture? _____

2. How did you identify the solids? _____
   _____
   _____
   _____

3. What happened to the water? _____

Cambridge Primary: Ready to Go Lessons for Science Stage 5 © Hodder & Stoughton Ltd 2013

Unit 1B: 5.2 Evaporation and condensation

Unit assessment

Questions to ask

- What happens when water evaporates?
- How is the process of condensation related to evaporation?
- What is temperature a measure of?
- What does a soluble solid mixed with water produce?
- What shape are salt crystals?

Summative assessment activities

Observe the learners while they participate in these activities. You will quickly be able to identify those who appear to be confident and those who may need additional support.

Evaporation or condensation?

This game assesses the learners' understanding of evaporation and condensation.

You will need:

A set of pre-prepared cards of pictures of everyday situations showing either evaporation or condensation, for example a kettle boiling, washing drying on a line, a steamed-up mirror, droplets of water on the outside of a glass containing an ice-cold drink.

What to do

- Organise the learners into a small group, to take turns around the table.
- Place the cards randomly face down on the table. Ask one learner to choose a card and say if it is an example of evaporation or condensation. They should then describe, using appropriate scientific vocabulary, exactly what is happening in the picture in terms of the process depicted.
- If they give a correct response, they should keep the card.
- Take turns until all the cards are used up.
- The winner is the learner with the most cards in their possession.

Taking temperatures

This activity assesses the learners' understanding of temperature.

You will need:

A range of different liquids, thermometers with different temperature ranges on them.

What to do

- Work with the learners either individually or in small groups.
- Give them the name of a liquid (for example water at room temperature, a hot cup of coffee, water in a washing machine) or an object (for example an ice cube, your body).
- Either ask them what the temperature is (water boils at 100°C and freezes at 0°C – human body temperature is 37°C) or ask them to show you the thermometer that they would choose to measure the temperature of the liquid or object, for example room temperature.
- As an additional activity, the learners could check their predictions using thermometers.

Written assessment

Distribute photocopiable page 65. The learners should work independently, or with the usual adult support they receive in class.

64

Name: _____

Making crystals

1. Imagine that you are visiting a planet in space. You find some blue crystals in a shallow pool and would like to bring some back to Earth.

2. Describe how you would use the liquid from the pool and make the biggest crystals that you could. Draw or write in the boxes below.

a) What would you do with the blue water?

b) What scientific process or processes would be involved?

c) Explain why your crystals would be large.

Cambridge Primary: Ready to Go Lessons for Science Stage 5 © Hodder & Stoughton Ltd 2013

Unit 2A: 5.3 The life cycle of a flowering plant

Flowers and fruits

Learning objectives

- Know that plants reproduce. (5Bp2)
- Make relevant observations. (5Eo1)

Resources

Internet access or reference books; examples (real or imitation) of plants with fruits or flowers on them; photocopiable page 67; magnifying glasses or hand lenses; a range of art materials – pencils, paint, collage materials, fabrics and threads; paper.

Starter

- Ask the learners to identify with talk partners as many of the fruits and flowers available as they can within a given time limit. Alternatively, present a series of pictures and organise this as a quiz, awarding a small prize for the winning individual, table or team.
- Discuss what the difference is between fruits and flowers. Alternatively, show pictures and ask the learners to identify them as fruits or flowers.
- Ask the learners to think about the purposes of each and to share their thoughts with the rest of the class.
- This will give you an insight into how much detail regarding flowering plants the learners have remembered from other work on plants in previous years.

Main activities

- Explain that flowers and fruits are the produce of flowering plants and that it is from these that new plants grow.
- Give out photocopiable page 67 and ask the learners to draw a storyboard to describe how they think this happens. Explain that a storyboard is a series of pictures showing the stages of plant growth (seed, small plant, growing plant, plant with fruit or flowers) – starting from when the seed is planted.
- Give the learners the opportunity to closely observe some actual flowers and fruits – use magnifying glasses or hand lenses if preferred (or available).
- Ask the learners to make a representation of the flower or fruit they have chosen or been given. Explain that this could be a drawing, painting, collage or textile work – depending on the range of art materials and time available. Ask them to make the piece an insect's view, to give a sense of enlarged perspective. The work could be 2D or 3D – it could also take more than one lesson to complete.

Plenary

- Ask some of the learners to share their storyboards with the rest of the class. Discuss their pictures to check their understanding.
- Remind the learners that flowers grow from seeds, that fruits contain seeds and that seeds are the basis of reproduction in flowering plants.

Success criteria

Ask the learners:

- Why do flowering plants have flowers or fruits?
- Where do flowers and fruits grow from?
- What happens to fruit when it grows?
- Why are fruits important for plants?

Ideas for differentiation

Support: Ask these learners to describe and draw the process in three pictures. Suggest ideas for their artwork or give them fewer alternatives to select from.

Extension: Ask these learners to choose and research the life cycle of a particular flowering plant native to their country.

Name: _____

Flowers and fruits

Draw and label how you think that new plants grow from a flower or fruit.

1.

Seed is planted.

2.

3.

4.

Unit 2A: 5.3 The life cycle of a flowering plant

Seeds

Learning objectives

- Know that plants reproduce. (5Bp2)
- Make relevant observations. (5Eo1)
- Present results in bar charts and line graphs. (5Eo4)

Resources

A selection of fruits with seeds, e.g. melon, lemon, orange, kiwi, pomegranate; plastic knives; paper plates or napkins; flipchart and markers; photocopiable page 69; card; paint, paint rollers; scissors; ribbon; glue; calendar tabs (if doing calendar activity).

Starter

- Give the learners in pairs a piece of fruit. Ask them to find the seed or seeds in the fruit – tell them that they may need to cut the fruit to find the seeds. Ensure good hygiene and hand washing before handling the fruit.
- Ask the learners to show the rest of the class their fruit and to describe where the seeds can be found, perhaps how many seeds there are, their colour and texture and if they make a pattern inside or on the fruit. This encourages good observational skills.
- Invite the learners to taste the fruit (be aware of any food intolerances). Perhaps they are trying a new fruit? Discuss their personal preferences.

Main activities

- Draw up a tally chart with assistance from the learners of their favourite fruit choices from the available fruits.
- Demonstrate how to use this information to construct a bar chart. Emphasise the need for a title, labelled axes, appropriate scale and presentation.
- Give out photocopiable page 69 for the learners to draw their own bar chart on. Ask those learners who need support to use the data already obtained and to re-draw the class bar chart. Invite all the other learners to collect a new set of data from all class members, asking them to name their favourite fruit – even if they have not tasted it in this lesson.
- Allow the learners to use more cut fruit to make some prints, by placing the cut side of the fruit into some paint and pressing it on a piece of paper. They could make a greetings card or calendar or a gift tag. Alternatively they could design and make their own gift wrap, printing a repeating pattern on a piece of paper.

Plenary

- Ask the learners questions relating to information contained in the bar chart.
- Invite them to show each other their printed products.
- Explain that seeds are necessary for reproduction to produce new flowers or fruits.

Success criteria

Ask the learners:

- Which fruit do we like best?
- How many more people prefer fruit A to fruit B?
- Why do plants have seeds?
- Invite the learners who need extension to ask their questions about the bar chart that they have constructed.

Ideas for differentiation

Support: Ask these learners to reproduce the class bar chart.

Extension: Ask these learners to write three questions about their bar chart to ask other learners when they show their completed bar chart.

Name: _____

Seeds: what is your favourite fruit?

1. Complete a tally chart for the responses of children in your class.

Fruit	Tally

2. Use the tally chart to draw a bar chart. Remember! Title, labels, scale.

Title: Bar chart to show _____

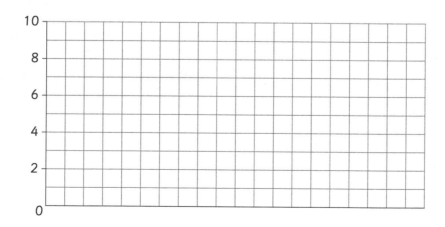

Cambridge Primary: Ready to Go Lessons for Science Stage 5 © Hodder & Stoughton Ltd 2013

Unit 2A: 5.3 The life cycle of a flowering plant

Planting seeds

Learning objectives

- Know that plants reproduce. (5Bp2)
- Identify factors that need to be taken into account in different contexts. (5Ep6)
- Discuss the need for repeated observations and measurements. (5Eo3)

Resources

Fruit seeds or stones, e.g. avocado, mango, lemon – one type only, or several different types; plastic cups; compost or growing medium; small hand trowels or forks; photocopiable page 71; examples of real or imitation plants or pictures – internet or books to show the fully grown fruit trees or bushes.

Starter

- Ask the learners if they can remember ever growing plants from seeds at school or at home before. (If they have studied the Cambridge Primary curriculum framework they should have.) Ask them to share their planting experiences with the rest of the class.
- Then ask if any of them have grown fruit seeds (*Ready to Go Lessons for Science Stage 1* page 160 gives details of how to grow an avocado seed – this would also be a suitable method for a mango seed).
- If necessary, demonstrate how to do this to remind the learners of some of their previous learning and to model for them how to plant seeds in this lesson.

Main activities

- Explain that in this lesson the learners will have the opportunity to plant some fruit seeds. All the learners could plant the same kind of seed or different groups could each plant a different type of seed, depending on the seeds available.
- Organise the class into appropriate groups and allow them to plant some fruit seeds.

- Discuss the main points to consider – depth of planting, amount of water, where to site the growing plant, and so on.
- Give out photocopiable page 71 for the learners to record what they have done.
- Show or look at the examples of fully grown fruit trees and bushes. Ask the learners to identify them and perhaps to guess the age of any real plants that are available. They need to realise that it can take many years of the correct nurturing before a fruit tree bears fruit.

Plenary

- Explain that seeds reproduce to grow into new plants of the same kind.
- Discuss how the learners might nurture the plants during their growth. Agree roles, for example for daily watering if considered necessary.

Success criteria

Ask the learners:

- How much water did you use? Why?
- How long do you think it will be before you can see signs of growth?
- What might you look for?
- Why is it a good idea to repeat measurements?

Ideas for differentiation

Support: Assist these learners during the planting process. Support them in completing photocopiable page 71. This could be done individually or as a group.

Extension: Ask these learners to predict which seeds might show signs of growth first and why.

Name: _____

Growing fruit seeds

I planted a _____ seed.

Method (what you did)

1. _____
2. _____
3. _____
4. _____
5. _____
6. _____

Diagram (how you set it up)

Unit 2A: 5.3 The life cycle of a flowering plant

Seed dispersal

Learning objectives

- Observe how seeds can be dispersed in a variety of ways. (5Bp3)
- Make relevant observations. (5Eo1)

Resources

Seeds planted in previous lesson; flipchart and markers; a large space, e.g. sports hall; PE / gym mats; a source of music and pre-selected tunes; photocopiable page 73; internet access or reference books.

Starter

- Look at the seeds planted in the previous lesson for any obvious signs of growth. Make this the initial activity in all subsequent Science lessons. Discuss and compare changes in growth over time. It is important for the learners to realise that not all fruit seeds will grow into fully mature plants.
- Take the class into the school hall or gym. Ensure that the learners are wearing appropriate footwear or are in bare feet, according to school rules. Place PE / gym mats randomly around the space. When the music plays, the learners can move freely around the hall, but when the music stops, they have to stand on a mat. Repeat this activity several times until they understand what they have to do.
- After several attempts, remove a mat each time. This will mean that they will need to crowd onto the remaining mats. Make a rule that if they are not standing with both feet on a mat after the music stops, or if they are the last person to stand on a mat, then they are 'out' and have to sit out of the game. The winner or group of winners are those remaining when all but one of the mats have been removed.
- Ask the learners to decide with talk partners what things green plants need to grow.
- Discuss their answers – clarify that all green plants need air, light, minerals and water.

Main activities

- From the Starter activity observations, introduce the idea that seeds from any plant are in competition with each other to find the best growing conditions. In the game the learners played, they were competing for space. Do seeds need to compete for space in the same way?
- Introduce the word 'dispersal', meaning the way in which seeds are scattered or spread out away from the parent plant.
- Give out photocopiable page 73 and describe how the learners need to look at the pictures of different seeds and decide how they might be dispersed.

Plenary

- Discuss why not being overcrowded is important for growing seeds – if they are too close to the parent plant, they will not survive. The parent plant is bigger and has a well-developed root system.

Success criteria

Ask the learners:

- What do all green plants need to grow well?
- What is the scientific word for scattering seeds?
- What do seeds dispersed by wind usually look like?
- Why do growing seeds compete for the things they need?

Ideas for differentiation

Support: Assist these learners when completing photocopiable page 73.

Extension: Ask these learners to find as many different examples of different methods of seed dispersal as they can.

Name: _____

Seed dispersal

Look at these seeds and think about how they might be dispersed.

Look at the shape, size and detail to help you decide.

Use some of these words to help you.

| hooks | juicy | light | spiky | wings |

coconut

jacaranda seed pod

castor oil fruit

flame of the forest tree seed pod

Unit 2A: 5.3 The life cycle of a flowering plant

Investigating seed dispersal

Learning objectives

- Observe how seeds can be dispersed in a variety of ways. (5Bp3)
- Make relevant observations. (5Eo1)

Resources

Internet access or pictures of seeds; magnifying glasses; a selection of different seeds; photocopiable pages 75 and 76; hairdryers or electric fans; measuring tapes or sticks.

Starter

- Show some examples from the internet or books of different methods of seed dispersal – wind, water, air, animal. Talk about the features of the different types of dispersal.
- If you have internet access, look at www.crickweb.co.uk/ks2science.html and try the seed-dispersal activity.
- Look at some different types of seeds and study their particular characteristics. Use magnifying glasses to observe the seeds more closely. Can the learners decide how the seeds might be carried or spread away from the parent plant, for example hooks, feathery seeds, size of seed?

Main activities

- Explain that in this lesson the learners will investigate the method of seed dispersal by wind.
- Ask the learners to discuss with talk partners what characteristics wind-dispersed seeds have. (They are light and often have feathery protrusions.)
- Ask: *Which seed will be dispersed furthest by wind?* Show the learners the selection of seeds available for use in the investigation.
- Explain that they will be given some different seeds and will need to predict and give a scientific reason why one seed will be dispersed the furthest distance. Different groups of learners could be given different seeds, or all groups could be given the same seeds to test, depending on what type and how many seeds are available.
- Working in pairs or small groups, the learners will need to decide what they will use as a source of wind (Electric fan, hairdryer, their own breath?) and how to make it a fair test – prompt them to consider where the seed will be dropped from and how they will measure the distance.
- Discuss possible ways of recording their results, for example a table or chart. Will they be able to use their results to draw a graph?
- Give them time to plan and carry out the investigation. Working in pairs or small groups, they should use photocopiable pages 75 and 76 to do this.
- Check their plans, giving appropriate guidance where necessary. Allow the learners to carry out their investigations once their plans have been checked. They need to be able to drop the seeds and measure the distance from the wind source.

Plenary

- Invite different groups of learners to share what they did and to discuss and explain their results and conclusions.
- Discuss the need for dispersal away from the parent plant as a means of survival.

Success criteria

Ask the learners:

- Was your prediction correct?
- How did you make the test fair?
- Which seed was dispersed the furthest?
- Why do seeds need to be dispersed away from the parent plant?

Ideas for differentiation

Support: Organise these learners to work in mixed-ability groups, or provide adult support for each group if they work together.

Extension: Challenge these learners to find as many different examples as possible of plants from around the world whose seeds are dispersed by the wind.

Name: _____

Investigating seed dispersal 1

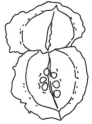

> Which seed will be dispersed the furthest?

Prediction

Predict which seed you think will be dispersed the furthest and say why.

I think that the _____ seed will be dispersed the furthest because
_____
_____.

Equipment

List the equipment you will need:

- _____
- _____
- _____
- _____
- _____
- _____
- _____

Method (what you did)

1. _____
2. _____
3. _____
4. _____
5. _____

Cambridge Primary: *Ready to Go Lessons for Science Stage 5* © Hodder & Stoughton Ltd 2013

Name: _____

Investigating seed dispersal 2

Which seed will be dispersed the furthest?

Results (what happened?)

Conclusion (what you found out)

Was your prediction correct? yes / no

Which seed was dispersed the furthest? _____

Explain why you think this happened.

_____

_____

_____

Unit 2A: 5.3 The life cycle of a flowering plant

Germination

Learning objectives

- Recognise that flowering plants have a life cycle including pollination, fertilisation, seed production, seed dispersal and germination. (5Bp7)
- Make predictions of what will happen based on scientific knowledge and understanding, and suggest and communicate how to test these. (5Ep3)
- Identify factors that need to be taken into account in different contexts. (5Ep6)

Resources

Plant pots or small containers for growing seeds in; potting compost; dried beans or peas; water; measuring jugs or cylinders; photocopiable pages 78 and 79.

Starter

- Explain to the learners that seeds in themselves are not growing. We say that they are dormant. Ask the learners to discuss with talk partners what the word 'dormant' means. Share your answers as a class.
- Explain that the first stage in the growth process of a seed is called germination. Germination is the next stage after seed dispersal. Explain that in this lesson the learners will think about: *What do seeds need in order to germinate?* (That is, begin to grow.)
- Ask the learners to discuss with talk partners conditions for germination. Share ideas as a class. (They will probably suggest light, water and warmth.)

Main activities

- Explain that different groups of learners will be given different situations to place some growing seeds in and that the results will be compared to find out the best conditions for germination.
- Decide what conditions you will test the seeds in, for example in the light or in the dark, with or without water and in cool or mild temperatures.
- Ask the learners how to make the test fair. The factors to be kept constant will need to be the type and number of seeds, the same size pot and the same amount of compost. If they agree that the seeds need water to grow, then the same amount of water should be poured into each seed-growing pot.
- Discuss and agree which conditions each group will establish. In pairs or small groups, allow the learners to plant up the beans or peas and place the pot in the agreed position and conditions.
- Give out photocopiable pages 78 and 79 for the learners to describe their method and to record their results.
- Discuss how to record the results in groups and as a class. Observe over at least a week. Record observations on a daily basis.

Plenary

- Explain that seeds have enough food already stored in them to germinate. The food stored in the seed is dry. It needs to dissolve before germination occurs, so seeds need water.
- Growth is affected by temperature. Seeds need the correct temperature for them to germinate; some need cold, others warmth.

Success criteria

Ask the learners:

- What kind of seeds did you plant?
- How did you make the test fair?
- Which conditions do you think will be best for germination?
- What do seeds definitely need to be able to germinate?

Ideas for differentiation

Support: Allow these learners to work in mixed-ability groups or in a group with adult support.

Extension: Ask these learners to predict which seeds will germinate first and why.

Name: _____

Germination 1

Prediction

I think that the seeds in the _____ (growing condition) will

germinate first because _____.

Method (what you did)

1. _____

2. _____

3. _____

4. _____

5. _____

Diagram (how you set it up)

Name: _____

Germination 2

Results (what happened?)

Write or draw to show what happened.

```
┌─────────────────────────────────────────────────┐
│                                                 │
│                                                 │
│                                                 │
│                                                 │
│                                                 │
│                                                 │
│                                                 │
└─────────────────────────────────────────────────┘
```

Conclusion (what you found out)

1. Which conditions made the seeds germinate the quickest?
    _____

2. How long did it take for signs of germination to be seen in the seeds that you planted?

    _____ days.

3. List two things that seeds definitely need to be able to germinate:

    a) _____

    b) _____

Unit 2A: 5.3 The life cycle of a flowering plant

Evidence of germination

Learning objectives

- Recognise that flowering plants have a life cycle including pollination, fertilisation, seed production, seed dispersal and germination. (5Bp7)
- Discuss the need for repeated observations and measurements. (5Eo3)
- Interpret data and think about whether it is sufficient to draw conclusions. (5Eo8)

Resources

Seeds planted in the previous lesson; internet access or pictures from books and a visualiser; interactive whiteboard or flipchart and markers; photocopiable pages 81 and 82.

Starter

- Use pictures from the internet or books showing different seeds from around the world and how they germinate, for example some trees and bushes in the Australian desert need fire to help them germinate. Coconuts only germinate after they have been soaked in salt water for some time. Other seeds only germinate after they have passed through the digestive system of an animal or bird.
- Give out photocopiable page 81 for the learners to complete and show their understanding of the conditions necessary for successful germination.
- Go through the answers together as a class, allowing the learners to mark their own work or that of their talk partner.

Main activities

- Ask the learners to think back with talk partners to the previous lesson and think again about the exact conditions necessary for germination. Discuss and agree with the rest of the class.
- Draw up a class list recording their thoughts. Display this prominently as a reference point for the lesson.

- Allow the learners to observe their seeds planted previously, record their observations and then to show or to comment and share their results and observations with the rest of the class.
- Discuss that once the seedlings begin to grow, they will need to be in the light. Green plants make their food using light energy. Allow the learners to choose the most suitable place in the classroom to move their growing seedling/s to.
- Give out photocopiable page 82 and explain that the learners need to use the information and interpret it. This will test some of their scientific skills, as well as their knowledge of what happens when seeds germinate.

Plenary

- Go through the answers to photocopiable page 82.
- Discuss and rectify any misconceptions and / or misunderstandings. (Some of the learners will think that light is necessary for germination – it isn't.)

Success criteria

Ask the learners:

- What grows first when a seed germinates? (The root.)
- What can you see develop next? (The shoot.)
- From where does the seed get its energy for this early growth? (Food is already stored inside it.)
- What do we call a young plant once it has germinated? (A seedling.)

Ideas for differentiation

Support: Provide adult support to help these learners complete photocopiable pages 81 and 82.

Extension: Ask these learners what would happen if the new seedlings were left in the dark.

Evidence of germination

Complete the sentences below. Use these words to help you (use each word once only).

| dark | food | root | seedling |
| shoot | warmth | water | |

1. In order to germinate, all seeds need _____.

2. Some seeds need light, others need to be left in the _____.

3. Different seeds need different temperatures, some need the cold, others need _____.

4. The thing that grows first after germination is the _____.

5. The next thing that grows is the _____.

6. To help the seed grow, it contains its own _____.

7. The young plant that begins to grow after germination is called a _____.

8. Write about a seed you know and describe the conditions it needs to germinate.

   Seed: _____

   It needs:

   _____
   _____
   _____
   _____

Name: _____

Germination observations

Shivi and Ahmed have put some seeds in a dish.

1. Why have they put the same number of each type of seed in the dish?

   _____

2. What did they give the seeds so that they would germinate?

   _____

Here are their results:

	Peas	Beans
Day 1	look a bit bigger	look the same as at the start
Day 2	roots appeared on 4 peas	no roots seen
Day 3	all 8 peas have roots	3 beans have roots
Day 4	the roots are growing longer	all 8 beans have started to grow roots
Day 5	first signs of shoots appear	no shoots seen

3. Which seeds began to grow first? _____

4. How can you tell?

   _____

   _____

5. Why did Shivi and Ahmed need to look at them every day?

   _____

Unit 2A: 5.3 The life cycle of a flowering plant

Insect pollination

Learning objectives

- Know that insects pollinate some flowers. (5Bp5)
- Recognise that flowering plants have a life cycle including pollination, fertilisation, seed production, seed dispersal and germination. (5Bp7)
- Make relevant observations. (5Eo1)

Resources

Internet access or pictures in books and a visualiser; face paints; yellow paper circles with double-sided sticky tape on both sides; photocopiable pages 84 and 85; coloured pens or pencils.

Starter

- Ask the learners to discuss with talk partners the question: *How are plants pollinated?* Discuss and share responses as a class.
- Explain to the learners that insect pollination is when insects transfer pollen to different parts of the same flower or to other flowers.
- Use the internet (if available) to show film clips of insect pollination. Enter the term 'insect pollination' into your search engine. Always check the film clips that you are going to use for suitability before the lesson.

Main activities

- Explain that insects pollinate flowers when pollen is rubbed off the insect's body. Ask the learners why insects visit flowers and what attracts them to flowers. (Some plants produce nectar, which is a sugary liquid, and many insects feed on this. Insects are attracted to flowers because of their scent, colours and nectar.) The insects carry pollen from flower to flower, and collect nectar for themselves. After pollination, the plant produces a seed.

- Ask for a volunteer to be a bee pretending to visit some flowers. Offer to paint the learner's face in black and yellow stripes to make them look like a bee (beware of any allergies). Ask for several other volunteers to be flowers. Give each flower a sticky circle to hold on the palm of each of their hands (pretending to be stamens). When the bee touches them, they stick their pollen onto the bee's body. When the bee visits another flower, it transfers some sticky pollen onto the new flower.

- Give photocopiable page 84 to the learners who need support and photocopiable page 85 to all the other learners. Before they complete it, explain that **ovum** is the singular of **ova**.

Plenary

- Invite some of the learners who need support to show the cartoon strips or storyboards they have produced showing what happens during pollination.
- Mention that insect pollination is not the only method of pollination. Some grasses and trees can be pollinated by wind. Humming birds pollinate some flowers and bats pollinate other flowers.

Success criteria

- What is pollen?
- Why are insects attracted to flowers?
- How do insects pollinate flowers?
- What happens in the plant after pollination?

Ideas for differentiation

Support: Give these learners photocopiable page 84. Ask them to draw a cartoon strip or storyboard showing what happens during pollination.

Extension: Ask these learners to research pollination in flowers from around the world, for example the Amazon water lily, or flowers pollinated by other methods such as wind or other animals.

Name: _____

Insect pollination

Draw a cartoon strip or storyboard to show what happens during pollination.

1.	2.
3.	4.
5.	6.

Name: _____

Insect pollination

Imagine that you are an insect. Write a diary extract for your day, explaining how you pollinated some flowers today.

Use drawings and illustrations if you wish.

_____

Now answer these questions.

1. How is an insect attracted to a flower?
   _____

2. Where is the pollen?
   _____

3. How does the insect transfer the pollen?
   _____

4. What happens when it reaches the ovum?
   _____

5. The process of male and female parts joining together to make a seed is called
   f_____.

Unit 2A: 5.3 The life cycle of a flowering plant

Plant structure

Learning objectives

- Observe that plants produce flowers which have male and female organs; seeds are formed when pollen from the male organ fertilises the ovum (female). (5Bp6)
- Know that plants reproduce. (5Bp2)
- Make relevant observations. (5Eo1)

Resources

A collection of real or synthetic flowers or pictures from the internet or books; hand lenses or magnifying glasses; photocopiable pages 87–89.

Starter

- Look at some examples of real or imitation flowers. If there are sufficient samples available, give the learners in pairs a flower to examine with their talk partner.
- Ask the learners to identify with talk partners the different parts of a flower.
- Share and discuss responses – prompt the learners with questions concerning the function of each flower part.

Main activities

- Explain that in this lesson the learners will look at and count the different parts of a flower. They might have a flower that is the same or different from that given to other learners.
- If there is time and specimens available outdoors to be picked, organise the learners to go outside and choose a flower for their group or pair to look at. Emphasise the need to only pick flowers where they have been given permission to do so. Be aware of any plant allergies in the group, or learners who might suffer from hay fever.
- Give out photocopiable page 87 to use as reference for the names of different parts of a flower. Explain to the learners who need support that they do not need to do the activity at the bottom of the page, but to complete photocopiable page 88 instead, using photocopiable page 87 to help them.

- Explain to all the other learners that they need to pull apart (gently) and count the different numbers of plant parts in their flower and then try to draw and label it on photocopiable page 87. Use hand lenses or magnifying glasses to aid observation.
- Ask all the learners to complete the table on photocopiable page 89.

Plenary

- Look at the completed diagrams on photocopiable pages 87 and 88.
- Discuss the completed tables on photocopiable page 89.
- Go to www.crickweb.co.uk/ks2science.html. Try the anatomy of a flower activity.

Success criteria

Ask the learners:
- What is the name of the flower you looked at?
- What is the function of the petals?
- What are the male parts of the flower called?
- What are the female parts?
- How many sepals did you find?
- Do all flowers have the same number of parts?

Ideas for differentiation

Support: Give these learners photocopiable page 87 for reference use only. Give them photocopiable page 88 to complete and label.

Extension: Give each of these learners, or each pair of learners, a different flower, and ask them to compare their findings at the end regarding numbers of petals, sepals, and so on.

Name: _____

Plant structure

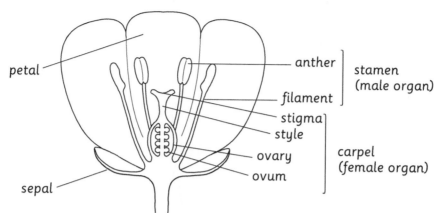

Draw a diagram of the flower you have been given and try to label it, naming as many parts as you can identify. Use the diagram above to help you.

My flower is a _____

Name: _____

Flower diagram

1. Look at the diagram on the 'Plant structure' page.
2. Use the following words to label the diagram below.

   ovary (female)   petal   sepal   stamen (male)

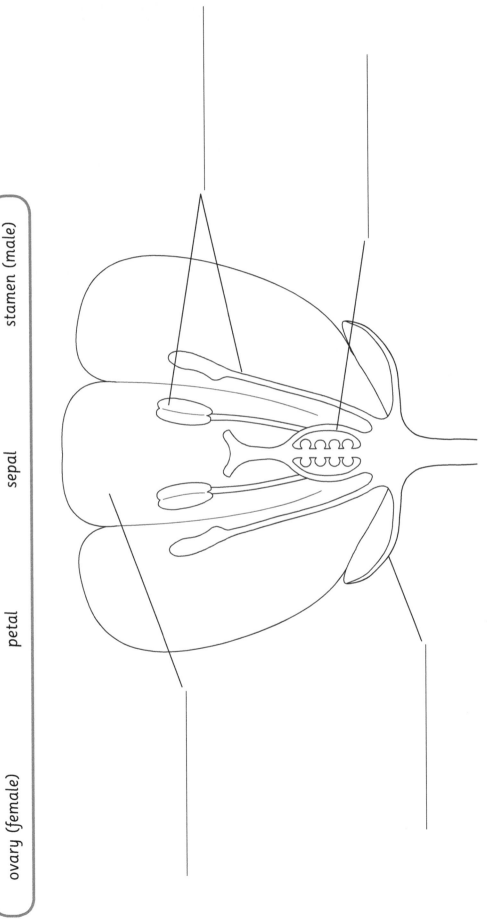

Name: _____

Structure of a flower

1. Name of flower: _____

2. Look at the flower you have been given or chosen.

3. Complete the table to show the different numbers of plant parts.

	How many?	Colour
Sepals		
Petals		
Stamens (male parts)		
Carpels (female parts)		

4. Now answer these questions.

   a) How many sepals does it have? _____

   b) What do sepals do? _____
   _____

   c) Does it have more male or female parts? _____

   d) Why? _____

Cambridge Primary: Ready to Go Lessons for Science Stage 5 © Hodder & Stoughton Ltd 2013

Unit 2A: 5.3 The life cycle of a flowering plant

Plant parts in pollination

Learning objectives

- Know that insects pollinate some flowers. (5Bp5)
- Observe that plants produce flowers which have male and female organs; seeds are formed when pollen from the male organ fertilises the ovum (female). (5Bp6)
- Make relevant observations. (5Eo1)

Resources

Yellow paper circles with double-sided sticky tape on both sides; face paints; real flowers with lots of pollen on the stamens; sticky tape; photocopiable pages 91 and 92.

Starter

- Play again the game from the lesson on page 83 that demonstrates pollination. Explain that in this lesson the learners will think in more detail about exactly what each plant part does in the process of pollination.
- As the pollination activity is acted out, stop the learners and ask at each point: *Which part of the flower does this happen in?* This will prompt discussion about the role of each plant part.
- Remind the learners that insect pollination is not the only method of pollination.
- Ask the learners if they have ever been covered in pollen, for example on their clothes or hands. Invite them to share their experiences of such.
- Lilies have lots of deep yellow pollen, which spreads easily. It can also stain quite badly, so some florists remove the stamens so that the pollen cannot stain clothes and soft furnishings.
- Give the learners a real flower and ask them to try to collect pollen from the stamens using sticky tape. Ensure they wash their hands carefully afterwards.

Main activities

- Following the Starter activities, ask the learners to discuss with talk partners exactly what happens to the pollen during pollination.
- Ask them to tell you again how insects are attracted to flowers. (Flowers attract insects by being colourful, large and pleasantly scented.)
- Remind the learners that insect pollination is when insects carry pollen from a plant's anthers to their stigmas, or to the stigmas of other flowers. Insects shake pollen out of the anthers and some lands on the sticky surface of the stigmas. Some pollen grains also stick to the hairy bodies or legs of insects. When they fly off, they carry the pollen with them to the next flower and so it is spread around.
- Give out photocopiable page 91 to the learners who need support. Explain that they have to read the statements then cut them out and stick them in the right order.
- Give out photocopiable page 92 to all the other learners and explain that they have to describe what happens in the different parts of a flower during pollination.

Plenary

- Invite the learners to show their work and discuss it.

Success criteria

Ask the learners:

- What are ova?
- What do ova need to grow into seeds?
- Where is pollen found?
- How does pollen get to the ova?
- What happens when the pollen meets the ova?

Ideas for differentiation

Support: Give these learners photocopiable page 91.

Extension: Ask these learners: *Do different flower colours and petal shapes attract different insects?*

Name: _____

Plant parts in pollination

Cut out the sentences from the bottom of the page. Stick them underneath the correct picture.

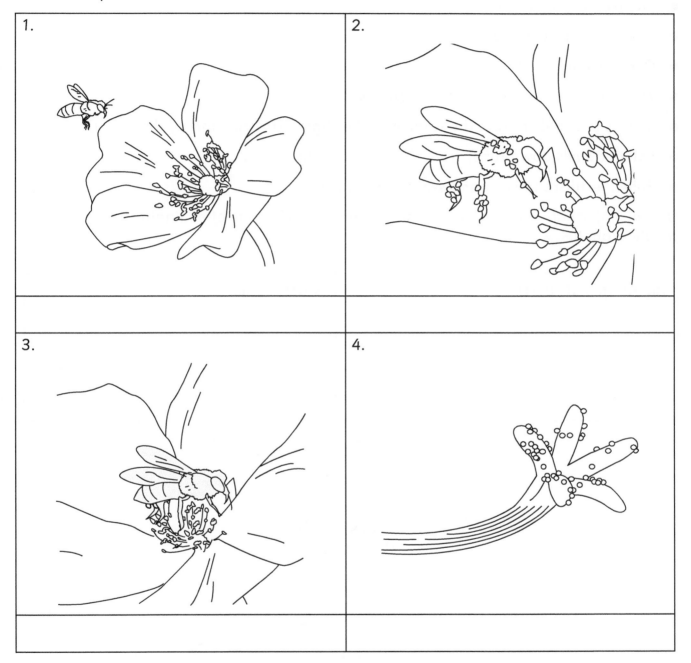

- Pollen sticks to the bee's hairy legs.
- The pollen gets to the ova.
- The bee moves to the stigma.
- The bee lands on the flower.

Name: _____

Plant parts in pollination

Pollination facts:
- The female part is the ovary.
- Ova grow inside the ovary.
- Ova need pollen to grow into seeds.
- The male part has pollen on its stamens.
- Insects take pollen from the stamen to the stigma (the top part of the ovary).
- Pollen travels down tubes from the stigma to the ovary.
- Pollen fertilises the ova and seeds are formed.

Use this information to describe what happens in each stage of pollination in the diagram below.

stigma (female part)	stamen (male part)

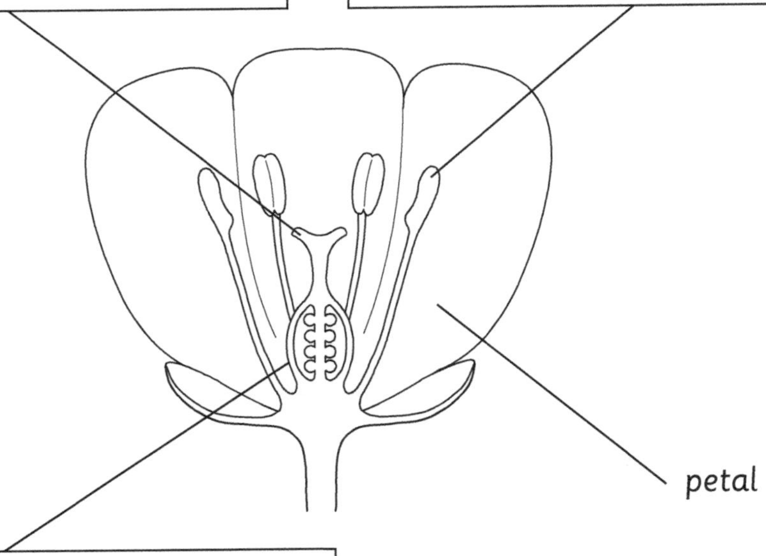

petal

ovary containing ova

Unit 2A: 5.3 The life cycle of a flowering plant

The life cycle of a flowering plant

Learning objectives

- Know that plants reproduce. (5Bp2)
- Recognise that flowering plants have a life cycle including pollination, fertilisation, seed production, seed dispersal and germination. (5Bp7)
- Make relevant observations. (5Eo1)

Resources

A small prize/s; photocopiable pages 94 and 95; sets of labels with the words / phrases describing each stage in the life cycle of a flowering plant – pollination, fertilisation, seed production, seed dispersal, germination, growth – enough for the number of groups you choose.

Starter

- Group the learners into two mixed-ability teams or a number of smaller teams.
- Give out photocopiable page 94 and set a time limit, for example 10 minutes, for the learners to complete their answers in their teams.
- Swap papers and give out the answers – award one mark for each correct response, and no half marks. The winning team is the group with the highest score. Check the winning score. Award a small prize to the winners. This could be in accordance with school policy on rewards, for example stickers, stars, house points.

Main activities

- Explain that in this lesson the learners will think about the life cycle of the flowering plant.
- Show the labels and ask the learners to tell you what happens at the stage written on the card.
- Give the learners in pairs or small groups a set of cards and ask them to arrange them in order (as a cycle), showing what happens in the life cycle of a flowering plant.
- Discuss and compare their answers and address any misconceptions or misunderstandings. Agree the order of the stages in the cycle.
- Give out photocopiable page 95 and explain that the learners have to label the stages on the diagram.

Plenary

- Go over the answers to photocopiable page 95.
- Explain that the life process is a cycle, which keeps going round and round. Seeds produce new flowers, which in turn make new seeds and so the plant survives. Compare this to the human life cycle: baby – child – adolescent – adult, and so on.
- Go to www.crickweb.co.uk/ks2science.html and try the life cycle of a flowering plant activity.

Success criteria

Ask the learners:

- How can you tell when a seed has germinated?
- When a (flowering) plant grows, what does it produce?
- How does pollination occur usually?
- What is the process when pollen and ova combine?
- What do all green plants need to grow well?

Ideas for differentiation

Support: Work with these learners to complete photocopiable page 95.

Extension: Ask these learners to write a sentence to explain what happens at each stage on the back of photocopiable page 95.

Team name: _____

Plant quiz

Answer these questions as a pair or team.

1. What do new plants grow from?
   _____

2. What is the scientific word for seeds being scattered?

   (There may be a bonus point for correct spelling!) _____

3. Name four different ways in which seeds might be scattered.

   (One point for each correct answer.)

   _____   _____

   _____   _____

4. What is the process when a seed begins to grow? _____

5. Which part of a new plant grows first? _____

6. What grows next? _____

7. What is pollination? _____
   _____

8. What do all green plants need to grow well?

   _____   _____

   _____

Name: _____

The life cycle of a flowering plant

Label the diagram with the correct terms.
Use these words to help you.

| germination | pollination | seed dispersal | growth |

seed production

fertilisation

Unit 2A: 5.3 The life cycle of a flowering plant

Unit assessment

Questions to ask

- Why do flowering plants have fruits or flowers?
- Why do plants have seeds?
- Name four different methods of seed dispersal.
- What happens during the process of pollination?
- Name the male parts of a flower.
- How do you know when germination has taken place?

Summative assessment activities

Observe the learners while they participate in these activities. You will quickly be able to identify those who appear to be confident and those who may need additional support.

Flowering plants

This activity assesses the learners' understanding of the life cycle of a flowering plant.

You will need:

A pre-prepared poster or diagram of the life cycle of a flowering plant (with no labels for stages); a set of labels containing each stage of the life cycle – pollination, fertilisation, seed production, seed dispersal and germination.

What to do

- Working with individual learners, present the diagram to them with different labels in place. Ask them to complete the cycle with the appropriate labels at each stage.
- For the learners who need extension, ask them to complete the whole cycle. Then ask these learners to describe what happens in detail at any stage that you ask them about.
- For the learners who need support, discuss the process on each label if they need help. If they find this too difficult, put all the labels except one on the chart and ask them about the process written on the missing label, discussing what comes before and after that stage in the cycle.

Seed dispersal

This activity assesses the learners' applied knowledge of the particular characteristics of seeds dispersed by wind / air, water, animals or explosion.

You will need:

A set of pictures showing seeds that have not been discussed in class.

What to do

- Ask the learners to suggest for each picture which method of seed dispersal each seed would use.
- Record their responses on a class checklist for future reference or report writing.

Written assessment

Distribute photocopiable page 97. The learners should work independently, or with the usual adult support they receive in class.

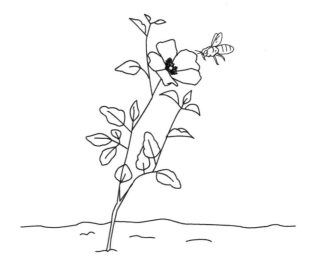

96

Name: _____

Germination

1. Complete the diagram to show what happens after germination has taken place.

2. Label the diagram.

3. What two things are essential for germination?

   _____

   _____

4. What is sometimes thought to be needed for germination, but in fact isn't needed?

   _____

Unit 2B: 5.4 Investigating plant growth

Conditions for germination

Learning objectives

- Investigate how seeds need water and warmth for germination, but not light. (5Bp4)
- Recognise and make predictions from patterns in data and suggest explanations using scientific knowledge and understanding. (5Eo7)

Resources

A copy of completed photocopiable page 95; internet access or reference books with diagrams and information about germination; identical plant pots or containers; cotton wool; water; bean seeds; access to a refrigerator; photocopiable page 99.

Starter

- Introduce this unit as building on work covered in Stage 3, and in Unit 2A: *5.3 The life cycle of the flowering plant* in this book.
- Ask the learners to think back to work in the previous unit on the life cycle of a flowering plant. Show photocopiable page 95 and revise the life cycle of a flowering plant, recapping the main events at each stage in the cycle, but not germination.
- Revisit the activity on the life cycle of a flowering plant at www.crickweb.co.uk/ks2science.html.
- Ask the learners to try to answer with talk partners the questions: *What does 'germination' mean?* and *What do seeds need to be able to germinate?*

Main activities

- Explain that in this lesson the learners will interpret data to show how much they understand about germination.
- With the learners, set up an experiment involving three identical pots with cotton wool in the base of each. Water the cotton wool. Place three bean seeds in each pot. Place the first pot (labelled A) in a warm position in the light. Place the second pot (labelled B) in the refrigerator. Place the third pot (labelled C) in a warm, dark cupboard.
- Discuss the different situations that the seeds are being exposed to for the purposes of this experiment. Explain that the learners will be able to observe any changes in these pots over the next few days.
- Ask the learners to make predictions about what will happen in each situation.
- Observe the pots daily over a school week. Record any observations – discuss as a class the best way(s) in which they might do this.
- Give out photocopiable page 99 and explain that this contains a set of results from a similar experiment. By answering the questions, the learners will be able to show how much they understand about the process of germination.
- Read through the page together and ensure that all the learners understand how to complete it.

Plenary

- Go through the learners' responses to photocopiable page 99. Explain any misunderstandings that might arise – some learners will still think that light is necessary for successful germination.

Success criteria

Ask the learners:

- What conditions did the beans in pot A have? (Warmth, light, water.)
- What was different for the beans in pot B?
- How did pots B and C differ?
- What is essential for germination? (Water and warmth, but **not** light.)

Ideas for differentiation

Support: Work with these learners to complete photocopiable page 99.

Extension: Set up a competition for these learners to grow the healthiest bean plant – let them decide the criteria for 'healthy' and allow them to plant beans and demonstrate their ideas.

Name: _____

Conditions for germination

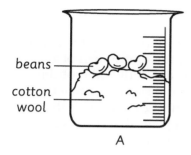
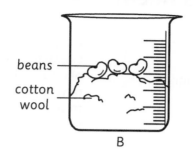
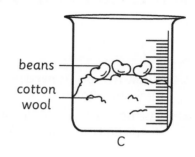

Method

- Pot A was placed in a warm, light position.
- Pot B was placed in a refrigerator.
- Pot C was placed in a warm, dark cupboard.

Results

Pot	Time taken for growth to be seen
A	5 days
B	no growth, even after 8 days
C	5 days

Now answer these questions.

1. What do the results for pot A tell us?

   _____

   _____

2. Why didn't the beans in pot B germinate?

   _____

   _____

3. What do the results for pot C tell us about conditions needed for germination?

   _____

   _____

Unit 2B: 5.4 Investigating plant growth

What do healthy plants need?

Learning objectives

- Know that plants need energy from light for growth. (5Bp1)
- Use observation and measurement to test predictions and make links. (5Ep2)

Resources

Flipchart and markers; internet access or pictures of different root systems in plants and trees; coloured water; celery stems or white flowers; vases or pots to hold the plants and water; measuring cylinders or beakers; photocopiable pages 101 and 102.

Starter

- Give out photocopiable page 101 and explain that the learners can use this to describe what a healthy plant needs to grow well.
- Invite the learners to share their responses. Ensure that they all know the main plant parts – roots, stem, leaf / leaves and flower or fruit. Discuss what green plants need to grow well – water, light and warmth.
- Explain that in this lesson the learners will think about how plants get water. In the following lessons, they will consider how plants get light and warmth.

Main activities

- Ask the learners: *How do plants get water?* It is important that the learners understand that one of the functions of the root system of a plant is to take up water (and minerals dissolved in it) from the soil. (Another function of the roots of a plant is to anchor it into the ground.) Do not allow use of the words 'suck' or 'drink' in describing this process as they are not scientifically correct. A plant does not suck or drink in the same way that a human can.

- Look at some specific examples of root systems. Compare, for example, the root system of a plant that grows in the desert (it will probably have a long tap root as it needs to get water from a long distance underground) with the root system of a garden plant that thrives in fertile soil (this will be shallow and branching). Discuss the similarities and differences, and possible reasons why.
- Ask the learners to demonstrate, using the plants available, how plants take up water. Give out photocopiable page 102 to help them plan this. Arrange the learners to work in small groups. Approve their plans before allowing them to set up their demonstration.

Plenary

- Invite groups of learners in turn to show the rest of the class what they have set up. Discuss their predictions and reasons why.

Success criteria

Ask the learners:

- What do all green plants need to grow well?
- What is the function of the root system?
- How do we know how water travels through the plant?
- Describe the differences in root systems between a garden plant and a desert plant.

Ideas for differentiation

Support: Work alongside these learners when they are completing photocopiable page 101. Allow them to work in mixed-ability groups for the practical activity.

Extension: Challenge these learners to measure how much water is taken up by a plant in 24 hours.

Name: _____

What do healthy plants need?

1. Draw a picture of a healthy plant and label the plant parts with the words below.

| flower | leaf | roots | stem |

2. Add any more useful plant words you know.

3. List three things a green plant needs to grow well.

   a) _____

   b) _____

   c) _____

Name: _____

How do plants take up water?

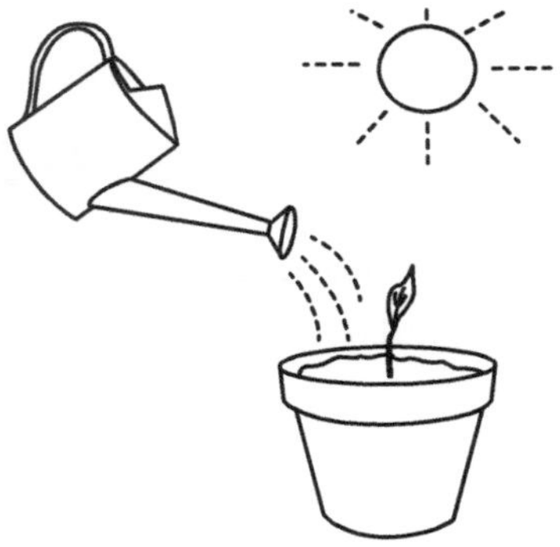

Design a test to show how plants take up water.

Equipment (list what you will need)

- _____

- _____

- _____

- _____

Diagram (draw a labelled diagram to show how you will set it up)

Predict (what do you expect will happen and why?)

_____

_____

Unit 2B: 5.4 Investigating plant growth

What do plants need light for?

Learning objectives

- Know that plants need energy from light for growth. (5Bp1)
- Make relevant observations. (5Eo1)

Resources

Plants in coloured water from the previous lesson; a selection of plants of the same type, but grown in different light intensities (dark, dim light conditions and normal daylight); flipchart and markers; photocopiable pages 104 and 105.

Starter

- Show the learners the list from the previous lesson of what they decided green plants need to grow well.
- Look at the plants from the previous lesson. Were the learners' predictions for taking up water correct? Invite them to show the rest of the class their results.
- Explain that in this lesson they will think about what plants need light for.

Main activities

- Show the learners the plants grown in different amounts of light. Ask them to describe each plant in turn. Discuss similarities and differences between plants that have been growing in the dark, in dim light conditions and in normal daylight. Give out photocopiable page 104 to the learners who need support and photocopiable page 105 to all the other learners.
- Look at and discuss their answers. Compile a word bank of good descriptive words used by the learners.
- Ask the learners to measure the heights of the individual plants grown in different light intensities. Record the results.

- Ask the learners how to improve the growth and appearance of the plants grown in dim light or in the dark. (They will probably suggest leaving them all in the light and keeping them watered.) Agree and decide as a class where to leave all the plants in order to observe them over the next week. Also agree how much water to give the plants and how regularly.
- The learners may also want to specify a particular time of day for watering the plants. Think about which equipment to use for measuring the amount of water and ensure that the learners agree that the test is fair.

Plenary

- Ask the learners to predict the growth of the plants over the next week. Ask them to write down their predictions. Discuss their reasoning. Explain that they will see if their predictions were correct in next week's lesson.
- Discuss that green plants need light to grow well and the Sun is the main source of light.

Success criteria

Ask the learners:

- How could you tell that a plant had been left in the dark?
- Describe the plant that had grown in normal daylight.
- Compare the plants grown in dim light and in the dark.
- Will all the plants be the same height next week?
- What do you expect the biggest changes will be in the plants that have been growing in low light conditions?
- How do green plants get light?

Ideas for differentiation

Support: Give these learners photocopiable page 104 to complete.

Extension: Ask these learners to construct a bar chart to show the growth of all three plants over the next week.

Name: _____

How do plants grow in different light?

1. Draw a picture in each box to show which plant you think was grown in which light conditions.

Dark	Dim light	Bright light

2. Look at the following word bank.

   a) Circle in black the words that describe the plant grown in the dark.

   b) Circle in blue the words that describe the plant grown in dim light.

   c) Circle in yellow the words that describe the plant grown in bright light.

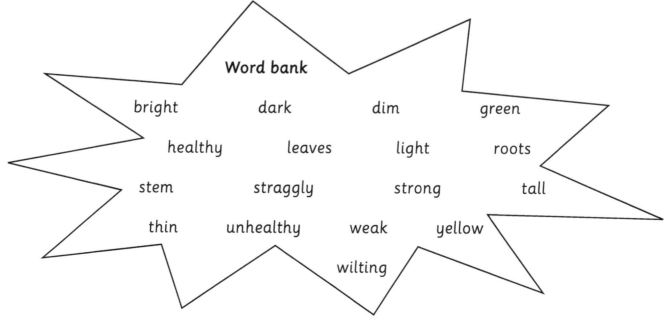

Word bank

bright   dark   dim   green

healthy   leaves   light   roots

stem   straggly   strong   tall

thin   unhealthy   weak   yellow

wilting

Name: _____

How do plants grow in different light?

Draw a labelled picture of each plant, describing the differences between them.

Dark	Dim light	Bright light
Description	Description	Description

Unit 2B: 5.4 Investigating plant growth

How does temperature affect plant growth?

Learning objectives

- Measure volume, temperature, time, length and force. (5Eo2)

Resources

Plants from the previous lesson; measuring sticks or tapes; several plants of the same type and similar sizes; water; measuring cylinders; thermometers; heat sources (if requested by the learners, e.g. electric lamps); photocopiable pages 107 and 108; different plants for extension work.

Starter

- Look at the plants left in the light from the last lesson. Measure the height of each of them and compare the differences.
- Ask the learners to discuss with talk partners their predictions from the last lesson. Were they correct? Do green plants need light to grow?
- Explain that so far in this unit of work the learners have thought about how plants get water and how important light is for growth. Today they will think about what plants need warmth for. Ask: *Where do plants get warmth from?* Confirm that the learners understand that the Sun is the main source of natural light and warmth for plants. However, they may be aware of plants being grown in hot houses and that artificial (electric) heaters and lamps may provide artificial light and warmth in such situations.

Main activities

- Ask: *Do plants grow better with or without warmth?* Discuss how the learners might find this out. Make sure they consider the concept of fair testing, for example the amount of water and light must be kept the same. The only factor that should change is the temperature each time. Note: the word 'factor' (not 'variable') is used until the end of Stage 6.

- Give out photocopiable pages 107 and 108 for the learners to plan their investigations on. Check their plans before allowing them to set up and carry out the practical activities.

Plenary

- Discuss and share ideas from different groups of learners about how they have set up their experiments.
- Ask them to predict what the outcome might be. Encourage them to give scientific reasons for their thinking.
- Discuss that there is an optimum growing temperature for different plants.

Success criteria

Ask the learners:
- How did you change the warmth of the plant's environment?
- How did you make it a fair test?
- What did you measure?
- What did you observe?
- What do plants need warmth for?

Ideas for differentiation

Support: Provide adult support for these learners and guide them through the planning process, using appropriate questioning.

Extension: Give these learners a different type of plant from the ones used by the rest of the class. Ask them to compare similarities and differences at the end of the experiment.

Name: _____

Do plants grow better with or without warmth? 1

Plan a test to find an answer to the question above.

Equipment (list what you think you might need)

- _____
- _____
- _____
- _____
- _____

Diagram (draw a diagram to show how to set it up)

Name: _____

Do plants grow better with or without warmth? 2

Method

Write a set of instructions about what to do so that someone else could repeat your experiment exactly.

1. _____
2. _____
3. _____
4. _____
5. _____
6. _____

Results (how will you show what happened?)

_____
_____
_____

Conclusion (what you found out)

Give a scientific reason for what happened.

Plants grow better with / without warmth. (Circle the correct answer.)

Scientific reason: I think that this is because _____

_____
_____
_____

Unit 2B: 5.4 Investigating plant growth

How do plants grow best – with or without water?

Learning objectives

- Know that plants need energy from light for growth. (5Bp1)
- Use knowledge and understanding to plan how to carry out a fair test. (5Ep4)
- Measure volume, temperature, time, length and force. (5Eo2)

Resources

A pair of near-identical plants – one that has been watered regularly and another that has not been watered regularly; a selection of plants of the same type and similar size; water; measuring cylinders or beakers; photocopiable pages 110 and 111.

Starter

- Look at and discuss the results from the previous experiment on how warmth affects plants' growth. Were the learners' predictions correct?
- Show the watered and unwatered plants. With talk partners, ask the learners to discuss what they think has caused the differences between the two plants. Listen to and comment on their suggestions.
- Explain that the learners' task today is to find out the best amount of water for a plant to grow well.

Main activities

- Remind the learners that the factor to be altered each time is the amount of water – everything else needs to remain the same for the test to be fair. Ideally the question to be answered needs one plant to be watered but not the other. Discuss and agree what needs to be kept the same – the type and size of plant, plant pot, amount of compost, position, temperature, and so on.
- Alternatively, suggest that groups, other than the learners who need support, combine so that they have more than two plants to test, which will give them more scope for using varying amounts of water.
- Introduce the idea of a control – a plant that is kept in optimum growing conditions for that plant and used as a comparison for the rest of the results.
- Organise the class into pairs or small groups, depending on the number of plants available for testing purposes. Give out photocopiable page 110 to the learners who need support and photocopiable page 111 to all the other learners to plan their fair test.
- Check their plans before allowing them to set up their experiments.

Plenary

- Invite groups of learners to explain to the rest of the class how they have planned and set up their experiment and how they are going to record the results over time.

Success criteria

Ask the learners:

- How many plants did you use?
- Did you have a control?
- How much water did you decide to give each time?
- What might happen if you gave a plant too much or too little water?
- Which plant thrived?

Ideas for differentiation

Support: Work with these learners and give them two plants only – one to water and one not to water.

Extension: Ask these learners to find out the least and greatest amounts of water that can be bad for a growing plant.

Name: _____

How do plants grow best – with or without water?

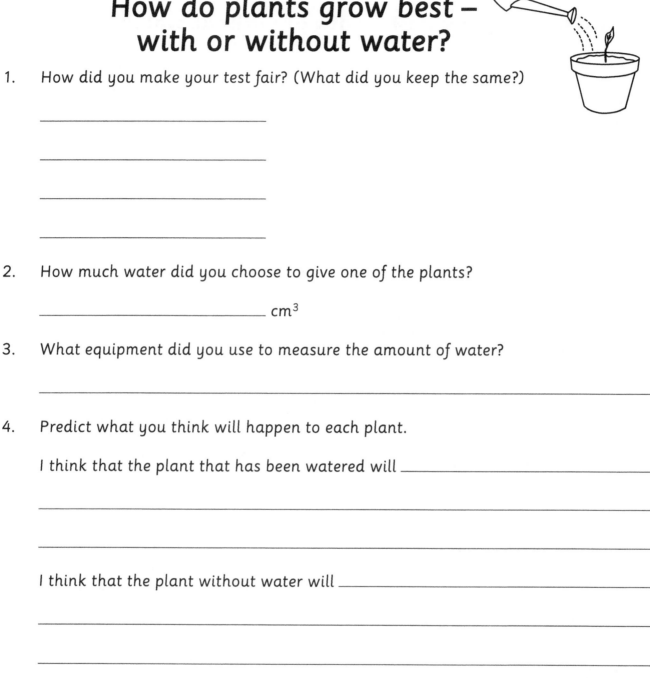

1. How did you make your test fair? (What did you keep the same?)

   _____

   _____

   _____

   _____

2. How much water did you choose to give one of the plants?

   _____ cm³

3. What equipment did you use to measure the amount of water?

   _____

4. Predict what you think will happen to each plant.

   I think that the plant that has been watered will _____

   _____

   _____

   I think that the plant without water will _____

   _____

   _____

   _____

Name: _____

How do plants grow best – with or without water?

1. How many plants will you use? _____

2. Are you using a plant as a control? yes / no

3. If so, what will you do to this plant during the experiment?

   _____
   _____
   _____

4. How much water will you give the plants?

Plant	Amount of water
control	
1	
2	
3	
4	
5	

5. What do you predict will happen?

   _____
   _____
   _____

6. When you have collected your results, complete this conclusion.

   The amount of water affects plant growth by _____

   _____

Cambridge Primary: Ready to Go Lessons for Science Stage 5 © Hodder & Stoughton Ltd 2013

Unit 2B: 5.4 Investigating plant growth

Photosynthesis

Learning objectives

- Know that plants need energy from light for growth. (5Bp1)
- Make relevant observations. (5Eo1)

Resources

A selection of different leaves; magnifying glasses, hand lenses and / or a microscope; wax crayons; photocopiable pages 113 and 114.

Starter

- Ask the learners to decide with talk partners what green plants can do that humans cannot do for themselves. (Make their own food.) Discuss the learners' responses and comment on any misconceptions.
- Explain that this makes every human and animal on the planet dependent on plants. We have to eat food to survive. We have to eat food (we cannot make it in our own bodies) and then digest it to help us move, grow and develop.
- Ask the learners which part of the plant they think that the food is produced in. (The leaves.)
- Give out a selection of different leaves. Allow the learners to use hand lenses, magnifying glasses and / or a microscope to look at them. Give out photocopiable page 113 for the learners to make a drawing on.
- Demonstrate how to make a leaf rubbing. (Place the photocopiable page on top of the back of the leaf (with the veins prominent). Then rub over the surface of the leaf through the paper with the side of a wax crayon to reveal the pattern of the veins.
- Discuss what the veins are and compare them to veins in the human body. (If you ask the learners to clench and unclench their fists several times, their own veins will become prominent on the back of their hands.)

Main activities

- Explain that the scientific word for the process by which green plants make their own food is 'photosynthesis' (from the Latin 'photo' meaning **light** and 'synthesis' meaning **building**). This means that plants can make food using the action of sunlight on the leaves. They use the energy from the Sun to make sugars, which are stored as starch. The leaves are the special parts of the plant for food production because they are green. This is because they contain a green pigment called chlorophyll and chlorophyll is needed for photosynthesis to take place.
- Give out photocopiable page 114 for the learners to complete, showing the process of photosynthesis.

Plenary

- Go over the equation for photosynthesis:
carbon dioxide + chlorophyll + water + light → starch (food) + oxygen
- Explain that the plant uses the food produced to help it grow.

Success criteria

Ask the learners:
- What four things do green plants need for photosynthesis to take place? (Carbon dioxide, chlorophyll, light and water.)
- Where do they get these things from?
- In which part of the plant does photosynthesis take place?
- How is the food that plants make stored?

Ideas for differentiation

Support: Provide these learners with adult support and guidance for completing photocopiable page 114.

Extension: Challenge these learners to find out if photosynthesis could take place in any other parts of a green plant.

Name: _____

Photosynthesis

You will need:
A leaf, a hand lens (or a magnifying glass or microscope), a wax crayon.

What to do

- Look at the leaf and draw a detailed diagram of it in pencil.

My _____ leaf

- Now place this page over the leaf and rub over it using the side of the wax crayon.

My leaf rubbing

I can see _____

Cambridge Primary: Ready to Go Lessons for Science Stage 5 © Hodder & Stoughton Ltd 2013

Name: _____

Photosynthesis

1. Use these words to label the diagram below.

| carbon dioxide | chlorophyll | light |
| oxygen | starch (food) | water |

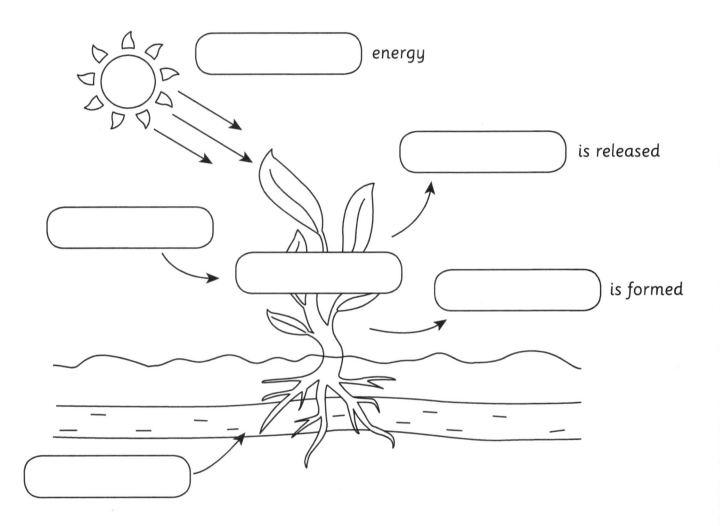

2. Complete this word equation for photosynthesis.

c _ _ _ _ _ _ d _ _ _ _ _ _ _ _ + c _ _ _ _ _ _ _ _ _ _ _ +
l _ _ _ _ _ + w _ _ _ _ → s _ _ _ _ _ _ (food) + o _ _ _ _ _

Unit 2B: 5.4 Investigating plant growth

Unit assessment

Questions to ask

- What do green plants need from light to be able to grow?
- What are the functions of the roots and root system in a flowering plant?
- How is water transported through a plant?
- What is important in controlling factors when designing a fair test?
- Can you give an example of using a control in an investigative situation?
- What is photosynthesis?

Summative assessment activities

Observe the learners while they participate in these activities. You will quickly be able to identify those who appear to be confident and those who may need additional support.

Growth conditions

This activity assesses the learners' understanding of the need for light for photosynthesis to take place.

You will need:

Two near-identical plants, one grown in the dark and the other in normal light conditions for that plant – preferably plants that are unfamiliar to the learners.

What to do

- Ask the learners to tell you why the plants have grown as they have.
- Record their responses.

Photosynthesis

This activity assesses the learners' understanding about the process of photosynthesis.

You will need:

A set of pre-prepared cards marked 'carbon dioxide', 'chlorophyll', 'water', 'light', 'starch (food)' and 'oxygen'.

What to do

- Set out the word equation using some of the cards for the process of photosynthesis.
- Ask the learners (individually if possible) to complete the equation using the rest of the cards.
- Ask them about what the reactants or products are from the cards they place.
- Ask them what photosynthesis is.
- The learners who need extension may be able to complete the whole equation unaided.
- Record their responses.

Written assessment

Distribute photocopiable page 116. The learners should work independently, or with the usual adult support they receive in class.

Name: _____

Germinating seeds

1. Describe and show how you would set up an experiment to show either that:

   a) Seeds need water for germination.

   OR

   b) Seeds need warmth for germination.

   OR

   c) Seeds **do not** need light for germination.

2. Choose which one you want to write about. Tick (✓) the list to show your choice.

3. Use the box to describe the experiment – draw or write what you would do. Use the back of this page also, if you need to.

   Remember! Diagram, method, results, conclusion.

Unit 3A: 5.5 Earth's movements

Earth, Sun and Moon

Learning objectives

- Explore, through modelling, that the Sun does not move; its *apparent* movement is caused by the Earth spinning on its axis. (5Pb1)
- Know that the Earth takes a year to orbit the Sun, spinning as it goes. (5Pb3)
- Make relevant observations. (5Eo1)

Resources

Flipchart and markers or whiteboard; photocopiable pages 118 and 119; internet access (if available); balloons; newspaper; glue; scissors; string; paints and painting equipment.

Starter

- Give out photocopiable page 118. Ask the learners to draw an annotated diagram of the Earth, Sun and Moon. (This is revision and will indicate to you how much they have remembered from previous work.)
- Invite some of the learners – maybe a learner who needs support, an average-ability learner and then a learner who needs extension work – to share and describe what they have drawn. Keep them in this order, asking each one to give more information than the previous learner.

Main activities

- This unit of work builds on and revises work covered in the Stage 2 units Light and dark and Day and night. Please be aware when introducing the Cambridge Primary Science Programme into school that the Stage 5 learners will not have covered this previously in Stage 2. It contains some difficult concepts.
- Give out photocopiable page 119. Ask the learners to complete it, giving facts that they already know about the Earth, Sun and Moon. Explain that this page can be added to over time as they study this unit.
- From the learners' responses to the Starter activity, compile a class list of Sun, Earth and Moon facts as an example. Write this as a table so that it can be displayed and the learners can add information to it throughout the unit. Use three columns, headed 'Sun facts', 'Earth facts' and 'Moon facts'. Display this prominently and in an accessible place for the learners. Explain that they can add information to this at any time – but they must read what is already there and not repeat anything! (This will begin to prepare them for some research work at the end of this unit.)

Plenary

- Show film clips of the Earth, Sun and Moon. Try www.bbc.co.uk/learningzone/clips/ and search Primary → Science → The Earth and beyond. Useful clips include 'The Earth's orbit around the Sun' (1:09), 'The Sun' (0:28) (the Sun and its distance from Earth) and 'The Moon and its orbit around Earth' (1:27).

Success criteria

Ask the learners:

- Which is the largest – Earth, Sun or Moon?
- How does the Earth move?
- How long does it take the Earth to rotate once on its own axis?
- How do we get a year on Earth?

Ideas for differentiation

Support: Pre-draw the Earth, Sun and Moon on photocopiable page 118 and ask these learners to simply label the diagram.

Extension: Challenge these learners to make models of the Sun, Earth and Moon (possibly to scale). Use papier mâché if there's time – layers of paper glued over an inflated balloon and left to dry before painting.

Name: _____

Earth, Sun and Moon

1. Draw a labelled picture showing the Earth, Sun and Moon in space. Think about how close they are to each other and how big each is.

2. Complete this table to describe how they move in space.

	How it moves
Sun	
Earth	
Moon	

Name: _____

Earth, Sun and Moon facts

1. Use this page to record what you already know about the Earth, Sun and Moon.
2. Add more information to it or correct your original ideas as you work through this unit.

Sun facts	Earth facts	Moon facts

Unit 3A: 5.5 Earth's movements

The Sun in the sky

Learning objectives

- Explore, through modelling, that the Sun does not move; its *apparent* movement is caused by the Earth spinning on its axis. (5Pb1)
- Make relevant observations. (5Eo1)
- Measure volume, temperature, time, length and force. (5Eo2)

Resources

Sticky tack; chalk; a large playground space that will be undisturbed all day; measuring sticks or tapes; photocopiable pages 121 and 122.

Starter

- Ask the learners to discuss with talk partners what the words 'sunrise' and 'sunset' mean. Discuss the learners' answers. Some of them will still think as they did in Stage 2 – that the Sun travels across the sky daily and somehow disappears when it is night-time. They imagine the Sun getting up in the morning and going to bed at night as they do.
- Using a small ball of sticky tack placed on a window, show the position of the Sun in the sky. Continue to do this at hourly intervals throughout the day, using a new ball of sticky tack each time.

Main activities

- Tell the learners that the experiment they are going to do in this lesson will help them to understand this apparent movement of the Sun better.
- Ask them to go outside and draw around their partner's shadow. Write the time on the shadow. Repeat this at half-hourly or hourly intervals throughout the day or morning. Make sure that the learners stand in exactly the same place each time their partner draws around their shadow.
- Give out photocopiable page 121 for them to record on the length of their shadow throughout the session or day.

- Discuss their measurements at the end of the session.
- Give out photocopiable page 122 for the learners to record their results on as a line graph.
- Demonstrate how this is done – give the graph a title, a good scale and label the axes.
- Ask the learners to draw their graph and answer the questions about it.

Plenary

- Discuss the learners' answers to the questions on photocopiable page 122.
- Explain that the Sun does not move – it is because the Earth spins on its own axis that the Sun **seems** to move.
- Look at the sticky tack on the window. What can you see? (A pattern in a curve.)

Success criteria

Ask the learners:
- When was your shadow the longest? (In the morning.)
- When was your shadow the shortest? (At noon.)
- What position in the sky was the Sun at both these times? (Low, then high.)
- Why did the length of your shadow change? (Because the Earth is turning.)
- What does the sticky tack on the window tell us?

Ideas for differentiation

Support: Assist these learners in measuring the length of their shadows. Work in a small group with them to draw their graphs.

Extension: Ask these learners to predict, then to find out, if the tallest learner has the longest shadow and the shortest learner has the shortest shadow.

Name: _____

Measuring shadows 1

Complete the table to show the length of your shadow at different times during the day.

Remember to stand in the same place each time.

Time	Length of shadow in cm

Name: _____

Measuring shadows 2

Use the graph paper below to draw a graph of your results. Remember to include a title and to label the axes. When you have drawn your graph, answer the questions below.

Title: Graph to show _____

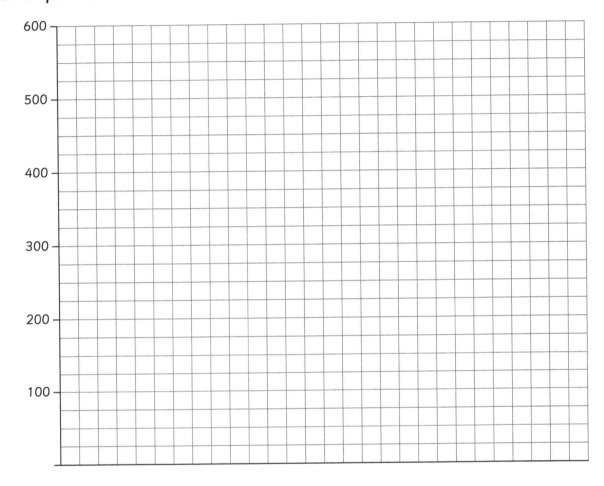

1. How long was your shadow at the start? _____ cm

2. How long was your shadow when it was the longest? _____ cm

3. What time was it when your shadow was the shortest? _____

4. What time was it when you had the longest shadow? _____

5. Why does your shadow change length? _____

_____

Cambridge Primary: Ready to Go Lessons for Science Stage 5 © Hodder & Stoughton Ltd 2013

Unit 3A: 5.5 Earth's movements

Day and night

Learning objectives

- Explore, through modelling, that the Sun does not move; its *apparent* movement is caused by the Earth spinning on its axis. (5Pb1)
- Know that the Earth spins on its axis once in every 24 hours. (5Pb2)
- Make relevant observations. (5Eo1)

Resources

A globe; a sticker or sticky tack; a torch; internet access; photocopiable pages 124 and 125.

Starter

- Show the globe and ask one of the learners to identify your country on it. Mark it with a sticker or small ball of sticky tack.
- Talk about which countries are close neighbours and which are on the other side of the world. Ask the learners about foreign countries that they might have visited or lived in.
- Ask one of the learners to demonstrate how the Earth turns. Introduce the word 'axis' and explain that the globe is orientated at the angle it is positioned in space, relative to the Sun.

Main activities

- Explain that the Earth makes one complete rotation or spin every 24 hours.
- Ask the learners what 24 hours makes. (One day.)
- Ask the learners to discuss with talk partners how, using a torch as the Sun, they could show how we have day and night on Earth, according to how the Earth rotates.
- Invite different pairs to demonstrate, using the globe and the torch, what they think happens to the Earth during the course of a day and a night.
- Go to www.bbc.co.uk/learningzone/clips/ and search Primary ⟶ Science ⟶ Earth and beyond. Use the film clips 'How does Earth's rotation create day and night?' (1:42) and 'How on Earth do we go from day to night?' (0:52).
- Give out photocopiable page 124 to the learners who need support and ask them to label and colour the diagram, showing how day and night occur on Earth.
- Give photocopiable page 125 to all the other learners. Ask them to draw and label a diagram showing how day and night occur and to answer the questions on the page.

Plenary

- Invite some of the learners to share their completed photocopiable pages.
- Either show the film clip again, or demonstrate using the torch and the globe – inviting the learners to provide the commentary as you do so.
- Add new information to the class chart 'Earth, Sun and Moon facts'. Allow the learners to add more information to their own pages.

Success criteria

Ask the learners:

- Name a country that has night when we have day.
- What do we call one turn of the Earth? (A rotation.)
- How long does it take for the Earth to make one full rotation? (24 hours.)
- What do 24 hours make? (A day and a night.)

Ideas for differentiation

Support: Give these learners photocopiable page 124 to complete.

Extension: Ask these learners to research more facts about planet Earth and add them to their 'Earth, Sun and Moon' fact sheets.

Name: _____

Day and night

1. Label and colour in the diagram. Use these words to help you.

   | day | Earth | night | Sun |

2. Complete this sentence.

   It takes _____ hours for the Earth to spin once on its axis to make a day and a night.

Name: _____

Day and night

1. How do we get day and night on Earth? Draw and label a diagram including the Earth and the Sun to show how this happens.

[ ]

2. Label the Earth and the Sun and show where it is day and where it is night.

3. What happens to the light from the Sun where it is daytime on Earth?
   _____

4. What happens to the Earth where it is night-time for us?
   _____

5. How long does it take for the Earth to rotate once on its axis?
   _____

Cambridge Primary: Ready to Go Lessons for Science Stage 5 © Hodder & Stoughton Ltd 2013

Unit 3A: 5.5 Earth's movements

How long is a day?

Learning objectives

- Know that the Earth spins on its axis once in every 24 hours. (5Pb2)
- Present results in bar charts and line graphs. (5Eo4)
- Recognise and make predictions from patterns in data and suggest explanations using scientific knowledge and understanding. (5Eo7)

Resources

Internet access, or diaries or local newspapers with sunrise and sunset times in them; clock faces; a map of your country; a map of a much larger country; photocopiable pages 127 and 128.

Starter

- Ask the learners what time it gets light in the morning at this time of year and what time it gets dark.
- Explain that in this lesson you will look at changes in day length in this country and throughout the world.
- Go to www.worldtime.com and select your country. Sunrise and sunset times will be available for several cities. Use these data to compare day length around your country. The map shows day and night at the time of viewing by the shadow on the map. Otherwise use diaries or newspapers to find this information.
- Give out photocopiable page 127 and ask the learners to complete individually or in groups as the data are found. Use clock faces to help calculate day length if necessary – depending on the learners' Maths ability and their familiarity and confidence in using a 24-hour clock. Ask them to rank the day lengths from longest to shortest.

Main activities

- Tell the learners that they will use the information about sunrise and sunset times to draw a bar chart and interpret the data.
- Use the data obtained for your own country to draw a bar chart – demonstrate this for the learners. Make sure that there is a title, correctly labelled axes and a correct scale. Give out photocopiable page 128 for the learners to draw the graph on. Allow the learners who need support to copy the graph for your country. Perhaps give the other learners data for one or more other countries and compare their findings once they have interpreted their bar charts.

Plenary

- Plot, as a class, the day-length times on a map of your country. Discuss what you find. (Cities on similar latitudes will have similar day lengths, cities on different latitudes will have different day lengths.)
- Explain that day length increases or decreases throughout the year, depending on the season.
- Invite some of the learners to plot the day-length times on maps of other countries (if used). Does a similar pattern occur?

Success criteria

Ask the learners:

- How do we calculate day length?
- Which city has the shortest day length?
- Which city has the longest day length?
- Why is that?

Ideas for differentiation

Support: Work in a small group with these learners to complete both photocopiable pages.

Extension: Ask these learners to find out what is special about the Land of the Midnight Sun or polar winters. (The Sun never rises.)

Name: _____

How long is a day?

Complete the table to show the day length for different cities in your chosen country.

Day length in cities in _____

City	Sunrise time	Sunset time	Day length	Order (1 = longest)

Name: _____

How long is a day?

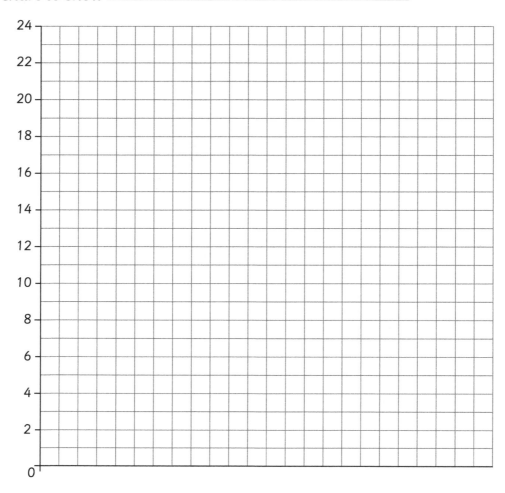

Title: Bar chart to show _____

1. Draw a bar chart to show day length across a country. Use the data you have collected or other data that your teacher gives you. Remember to include a title and to label the axes.

2. Which city / cities have the shortest day? _____
   _____

3. Which city / cities have the longest day? _____
   _____

4. Why? _____
   _____

Unit 3A: 5.5 Earth's movements

What makes a year?

Learning objectives

- Know that the Earth takes a year to orbit the Sun, spinning as it goes. (5Pb3)
- Make relevant observations. (5Eo1)
- Present results in bar charts and line graphs. (5Eo4)

Resources

A picture of the Solar System from the internet or books; a pair of sunglasses or a paper Sun; a globe; photocopiable pages 130 and 131.

Starter

- Show a picture of the Solar System with the Sun at the centre. Explain to the learners that they will find out more about different planets later in this unit.
- Describe how all the planets in our Solar System revolve around the Sun. Each complete journey around the Sun is called an orbit.
- Ask the learners to discuss with talk partners: *How does the Earth move in two different ways at the same time?* (It spins on its axis and orbits the Sun.)
- Demonstrate this using a learner as the Sun. This learner could wear sunglasses (and a big smile), or you could prepare a paper Sun for them to hold up. Let the 'Sun' stand still, then ask another learner to use the globe to demonstrate the Earth spinning on its axis and orbiting the Sun at the same time.

Main activities

- Look at the globe again and talk about the northern and southern hemispheres.
- Repeat the Starter activity, this time looking at each hemisphere as the Earth rotates and orbits the Sun. Explain that because the axis of the Earth is tilted, as it revolves around the Sun either the northern or southern hemisphere is tilted towards the Sun. This means that for people living in the hemisphere tilted towards the Sun there are more hours of sunlight and higher temperatures – it is summer. When one hemisphere has summer, the other hemisphere has winter because that half of the Earth is tilted away from the Sun.
- Give out photocopiable page 130 to the learners who need support, for them to draw arrows on the diagram indicating the two different ways in which the Earth moves in space.
- Give out photocopiable page 131 to all the other learners. Explain that they have to complete the diagram and answer the questions about how we get summer and winter.

Plenary

- Remember that one rotation of the Earth on its own axis takes 24 hours or one day. $365\frac{1}{4}$ of these rotations around the Sun = a year.
- Explain leap years: every four years there is an extra day in the year to account for the $4 \times \frac{1}{4}$ days.

Success criteria

Ask the learners:

- How many revolutions of the Sun does the Earth do each year? (One.)
- How many days does it take for the Earth to orbit the Sun? ($365\frac{1}{4}$)
- How many days are there in a leap year?
- Why do we have leap years?

Ideas for differentiation

Support: Give these learners photocopiable page 130.

Extension: Ask these learners to look for examples from around the world of bad weather conditions wreaking havoc.

Name: _____

What makes a year?

1. Draw arrows on the diagram to show the **two** different ways in which the Earth moves.

2. Complete these sentences.

    a) It takes __ __ __ days for the Earth to orbit the Sun.

    b) We call this a __ __ __ __.

Name: _____

What makes a year?

1. Draw arrows on the diagram to show the **two** different ways in which the Earth moves.

2. Which hemisphere is nearest the Sun? _____

3. What season will it be in this hemisphere? _____

4. How do summer and winter occur?

   _____

   _____

   _____

Cambridge Primary: Ready to Go Lessons for Science Stage 5 © Hodder & Stoughton Ltd 2013

Unit 3A: 5.5 Earth's movements

The Solar System

Learning objectives

- Research the lives and discoveries of scientists who explored the solar system and stars. (5Pb4)
- Make relevant observations. (5Eo1)

Resources

Flipchart and markers; internet access or reference books; photocopiable pages 133, 134, 135 and 136; internet software package for design; paper; paints; colouring pencils; letter stencils.

Starter

- As a class, thought-shower what the learners already know about space. Produce a mind map for general display. Allow the learners to contribute to this when they feel that they have something more to add.
- Use a search engine to find 'the Solar System' online. Choose a film clip that shows the Earth's Solar System of Earth, Sun, Moon and surrounding planets. There are many examples available – choose the one you like best.

Main activities

- Explain to the learners that in 2006 the planet Pluto was reclassified as a dwarf planet. This means that it is now considered too small to be classed as a major planet, unlike all the other planets in Earth's Solar System. Explain that it may still appear in some textbooks and on the internet classified as a minor planet, along with the other planets in our Solar System.
- If available, use the website chosen for the Starter activity to guide the learners through a systematic look at the planets in our Solar System. Think about their names, sizes, distances from the Sun and orbits. If there is no internet access, use reference books to find this information.
- Give groups of learners one planet each to research.
    - Give out photocopiable page 133 so that the learners can label a diagram of the Solar System.
- Distribute the information on each of the planets from photocopiable pages 134 and 135 to the groups according to which planet they are researching. Also provide reference books and internet access.
- Give out photocopiable page 136 for the learners to record their information on. This will be good preparation for their work at the end of this unit, which will be individual research.

Plenary

- Check the learners' completed photocopiable page 133. Ask relevant questions about the names, positions and relative sizes of the planets.
- Invite different groups, in turn, to share their planet findings with the rest of the class.
- Ask each group to produce an information poster about their planet, on paper or on the computer. Make a classroom display of the completed posters. This may take more than one lesson.

Success criteria

Ask the learners:

- How many planets are there in our Solar System?
- Which planet is nearest the Sun?
- Which planet is furthest away from the Sun?
- Name the planets in order, from nearest the Sun to furthest away.
- Name a famous scientist who researched space and the planets.

Ideas for differentiation

Support: Give these learners planet Earth to research to increase their knowledge about something that they already know some facts about.

Extension: Challenge these learners to find out the name of a famous scientist who discovered their planet or something about it.

Name: _____

The Solar System

Name and colour the planets in our Solar System.

Use these words to help you.

| Earth | Jupiter | Mars | Mercury | Neptune | Saturn | Uranus | Venus |

Cambridge Primary: Ready to Go Lessons for Science Stage 5 © Hodder & Stoughton Ltd 2013

The Solar System – information about planets 1

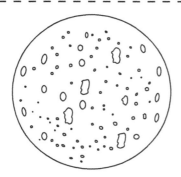

Mercury

Mercury is the nearest planet to the Sun.

It is 57 million km away from it.

Its maximum temperature is 420°C.

Its minimum temperature is −220°C.

It has no moons.

Venus

Venus is the second planet away from the Sun.

It is 108 million km away from it.

Its average temperature is 464°C.

It is the hottest of all the planets.

It has no moons.

Earth

The Earth is the third planet away from the Sun.

It is 150 million km away from it.

Its average temperature is 7.2°C.

It is the biggest of the planets in the inner Solar System.

It has one moon called Luna.

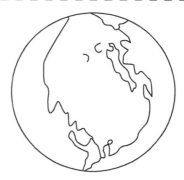

Mars

Mars is the fourth planet away from the Sun.

It is 228 million km away from it.

Its maximum temperature is 36°C.

Its minimum temperature is −123°C.

It has two moons called Deimos and Phobos.

The Solar System – information about planets 2

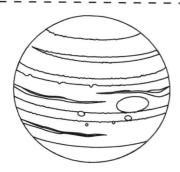

Jupiter

Jupiter is the fifth planet away from the Sun.

It is 779 million km away from it.

Its average temperature is −153°C.

It has a Great Red Spot.

It has 62 moons. Galileo discovered the four largest of these in 1610.

Saturn

Saturn is the sixth planet away from the Sun.

It is 1430 million km away from it.

Its maximum temperature is −184°C.

It has several hundred rings.

It has 53 moons. The best-known is called Titan.

Uranus

Uranus is the seventh planet away from the Sun.

It is 2880 million km away from it.

Its maximum temperature is −184°C.

It spins on its side.

It has 21 moons. The largest is called Titania.

Neptune

Neptune is the furthest planet away from the Sun.

It is 4500 million km away from it.

Its maximum temperature is −223°C.

It has a Great Dark Spot.

It was discovered in 1846 by Johann Galle and Heinrich D'Arrest.

Cambridge Primary: Ready to Go Lessons for Science Stage 5 © Hodder & Stoughton Ltd 2013

Name: _____

The Solar System – fact file

Use the information you have been given and any other information you can find from the internet or books to make a fact file for a planet.

Name of planet: _____

Picture of planet:

Distance from the Sun: _____

Number of moons: _____

Temperature: _____

Size: _____

Who discovered it? _____

When? _____

Interesting fact(s): _____

_____

_____

_____

Unit 3A: 5.5 Earth's movements

Space scientist – Aristarchos

Learning objectives

- Research the lives and discoveries of scientists who explored the solar system and stars. (5Pb4)
- Know that scientists have combined evidence with creative thinking to suggest new ideas and explanations for phenomena. (5Ep1)

Resources

Internet access or reference books; photocopiable pages 138 and 139.

Starter

- Explain that over the next few lessons you will share with the learners some facts about famous space scientists throughout history. The learners will then be able to choose to research one they are particularly interested in. Alternatively, depending on the availability of resources, you may tell them the name of the scientist you would like them to research.
- Remind the learners that they have begun to do some research in previous lessons, for example on the planets. Explain that you will give them some suggestions and help them to collect the information together and present it in an interesting way.
- Ask the learners to discuss with talk partners ways in which they could present the information they find out. Listen to their responses and suggest further ideas. Examples could include: an information leaflet (on the computer or by hand), a presentation (either using a PowerPoint presentation or giving a talk to the rest of the class using notes and props). Give other suggestions, such as collecting information to share with a younger class, producing a poster, writing a biography, making a model – the possibilities are endless, and exciting.
- Decide how much class time you will allocate to this task and maybe include some homework time.

Main activities

- Explain that during the next few lessons you will model some good ways to gather information for research.
- Discuss the danger of simply cutting and pasting information directly from the internet. Is it fit for purpose?
- Begin by giving the learners some time to look at any information available in class – from the internet or books – to consider who they might like to find out more about, or give them a name to research.
- As a class, think about suggestions for a front cover for a booklet or general information about the person being researched.
- Give out photocopiable page 139 to assist the learners with collating this information.
- Use the example of Aristarchos and the information given on photocopiable page 138 to demonstrate what photocopiable page 139 might look like once completed. Discuss or show how to work out the age of someone born BCE.

Plenary

- Ask each learner which space scientist they are going to research.
- Write a list of names and their subject.
- Discuss improvements and / or better ways to use the information collected for the purposes of photocopiable page 139.

Success criteria

Ask the learners:

- Which space scientist will you research?
- What do you need to avoid when using information straight from the internet?
- How will you present your finished research?
- Who will you make the information available to?

Ideas for differentiation

Support: Give these learners a smaller selection of space scientists for whom you have more information available to choose from.

Extension: Give these learners a more extensive list to choose from – or ask them to suggest other space scientists of their own.

Aristarchos – fact file

Place of birth: Samos, Greece (find a map of where in the world this is)

Born: Approx. 310 BCE

Died: Approx. 230 BCE

Famous: Mathematician – then astronomer

What he did: He was the first person to say that the Earth is not the centre of the universe. He said that it rotates and that it rotates daily on its own axis and yearly around the Sun.

Before this people believed that the Earth was the centre of the universe.

What he wrote: A book about the distances from Earth and the sizes of the Moon and the Sun. (These have now been found to be greatly underestimated.)

What he invented: A bowl-shaped sundial with a pointer inside it to cast a shadow in the middle of the bowl.

(Other scientists you may research could be Pythagoras, Copernicus, Galileo, Newton, Einstein, your own country's cosmonauts.)

Name: _____

Researching a famous scientist – fact file

Picture (drawing or photo):

Place of birth: _____

Date of birth: _____

Date of death: _____

Age: _____

Famous for what discovery or invention?

_____

_____

_____

_____

Cambridge Primary: Ready to Go Lessons for Science Stage 5 © Hodder & Stoughton Ltd 2013

Unit 3A: 5.5 Earth's movements

Space scientist – Pythagoras

Learning objectives

- Research the lives and discoveries of scientists who explored the solar system and stars. (5Pb4)
- Know that scientists have combined evidence with creative thinking to suggest new ideas and explanations for phenomena. (5Ep1)

Resources

Internet access or reference books; photocopiable pages 141–143.

Starter

- Ask the learners to share one fact about the scientist that they are researching.
- Ask for any other facts that they may have discovered about Aristarchos from the previous lesson.
- Discuss how they could use the information gathered. They could simply included it in a fact file about the person. Alternatively, they could use it to write a summary for the back page of a booklet or leaflet of information.
- Give the learners time to make notes for these purposes. Discourage copying completely. Help them to select relevant points and consider how to re-arrange the page for a balance of information and an easy reading style, for example using bullet points.

Main activities

- Explain that in this lesson you are going to share some information about Pythagoras.
- Ask the learners to discuss with talk partners: *Who was Pythagoras?*
- Listen and respond to their answers.
- Consider how they might use this information as part of their research – or how to find and use similar information about the person they are finding out about. For the purposes of a written presentation, ask the learners for suggestions for a front cover for their work.
- Share the information about Pythagoras from photocopiable page 141. Discuss the information. Did any of the learners know this already? What is the most interesting piece of information that you have shared with them? Explain that different information will appeal to different audiences.
- Explain to the learners that they are now going to design a front cover for their research booklet. Give out photocopiable page 142 as an example of a layout that could be used for a front cover. Discuss how to create a good title and make the cover appealing with an illustration (of the person or one of their inventions or discoveries) so that others will want to read the contents. How could the learners use the information about Pythagoras to do this? Model an example front cover, designed via discussion with the class.

Plenary

- Give the learners time to do more research or to share information with each other.
- Explain that they can provide similar information about their scientist in the same way.

Success criteria

Ask the learners:

- Who was Pythagoras?
- Where did he come from?
- What is he famous for in space exploration?
- What legend is associated with him?

Ideas for differentiation

Support: Assist these learners in designing a front cover for their work, using photocopiable page 142.

Extension: Provide these learners with photocopiable page 143 on which they can design a front cover and write some interesting 'blurb' for the back cover of their booklet.

Pythagoras – fact file

Born: Greek island of Samos, approximately 569 BCE

Died: Approximately 475 BCE

Famous for: Pythagoras' theorem in Maths

Scientific discoveries / theories: He was one of the first people to recognise that the Earth is a sphere in the centre of our universe.

He also noticed that the Moon's orbit is tilted towards the Equator of the Earth.

He recognised that Venus as an evening star is the same planet as Venus as a morning star.

Legend

This story claims that Pythagoras said that he could write on the Moon. He is said to have written on a mirror in blood and placed it opposite the Moon. The writing supposedly appeared reflected on the Moon's disc.

Name: _____

Front cover

by _____

Name: _____

Book cover

By _____

About this book

About the author

Unit 3A: 5.5 Earth's movements

Space scientists – Copernicus and Galileo

Learning objectives

- Research the lives and discoveries of scientists who explored the solar system and stars. (5Pb4)
- Know that scientists have combined evidence with creative thinking to suggest new ideas and explanations for phenomena. (5Ep1)

Resources

Internet access or reference books; photocopiable page 145.

Starter

- Again, start the lesson by inviting some of the learners to share their research – notes or finished pages. This will provide the other learners with good ideas.
- Invite the learners to comment on what they particularly like about a piece of work.
- Consider any ways in which a piece of work could be made even better. Make only positive criticism – use words such as 'it would be even better if …'
- Ask the learners to share any more facts that they might have found out since the last lesson about Aristarchos or Pythagoras. This will be useful for those studying these particular scientists.

Main activities

- Explain that today's lesson will be about two other famous scientists – Copernicus and Galileo. This will include watching a film clip and taking notes.
- Discuss the process of note-taking and how the learners do not need to write down every word – just important words that will help them to remember.
- Explain that you are also going to consider what would be a good order in which to present information and how to compile a contents list.
- Ask the learners if any of them know anything about Copernicus already.
- Go to www.bbc.co.uk/learningzone/clips/ and search Primary ⟶ Science ⟶ Earth and beyond ⟶ Copernicus and Galileo 'The Movement of the Earth' (4:58). This film clip is in the form of a game show that describes how Copernicus and Galileo discovered and explained the movement of the Earth around the Sun.
- Tell the learners that after the film clip you will ask them some questions about what they have found out. Ensure that they make notes either in their Science notebooks or on loose paper.
- Ask questions after viewing the film clip. Listen to the learners' responses and discuss any misconceptions. Write up a list of useful spellings for reference.
- Give out photocopiable page 145. Explain that this is one way in which the learners could begin to compile a contents list for their information about their chosen scientist. Discuss what they might include. Suggest such things as the scientist's fact file, a biography, a map of their birthplace, their greatest invention or discovery, legends about them, and so on.

Plenary

- Ask for a 'hands-up' signal from the learners – which of them will find working in the way illustrated on photocopiable page 145 useful or helpful?
- Assign time for them to begin their contents list, or for planning.

Success criteria

Ask the learners:

- Who was Copernicus and what was his main discovery?
- What is Galileo famous for?
- What will you include in your contents list?

Ideas for differentiation

Support: Help these learners with note-taking, or take notes for them to use.

Extension: Ask these learners to work on a presentation similar to the film clip involving two other scientists.

Name: _____

Famous scientist: _____

Portrait

Contents list

Title	Page

Unit 3A: 5.5 Earth's movements

Researching space scientists

Learning objectives

- Research the lives and discoveries of scientists who explored the solar system and stars. (5Pb4)
- Know that scientists have combined evidence with creative thinking to suggest new ideas and explanations for phenomena. (5Ep1)

Resources

Internet access or reference books; photocopiable pages 147–152.

Starter

- Explain that the remaining time in this unit will be given to individual research concerning the chosen or given scientists.
- Make the learners aware that you will be available to help, support and guide them with this work.
- Show them the in-school resources that are available and ask them to bring in useful information that they find at home to share with the other learners. Decide if materials can be borrowed and how long they can be kept for, to allow everyone a fair chance of using the resources available.
- Set the expectation that you would like one finished piece of work per lesson from each learner. This could be writing, a piece of artwork or a model, for example.

Main activities

- In each lesson, invite the sharing of new facts and / or information between the learners.
- Give opportunities for the learners to show pieces of their research to the rest of the class, for inspiration and improvement.
- Depending on the availability of resources, you might be able to organise the learners into groups using their own materials, photocopiable pages 147–152, school reference books, the school library and / or the internet. Rotate these groups accordingly throughout the rest of the sessions for this unit so that they all have opportunities to see and use all the resources available. The photocopiable pages provide basic reference material on a range of scientists from around the world throughout history.
- Work with individual learners and groups as necessary, asking relevant questions and challenging the more-able learners.
- Provide extra materials for them to use as they become available.
- Use the time in the lesson as you visit each group or individual to mark individual pieces of work to avoid having to mark lots of written work when the presentations are complete.

Plenary

- At the end of each lesson, allow time for the learners to showcase elements of their research.
- Find out what extra materials – especially art or model-making materials – might be required for the next lesson so that you can have them readily available for the learners to use.
- Set a final submission date for this piece of work. Remind the learners of this often.

Success criteria

Ask the learners:

- Tell me one fact about your scientist.
- What does your poster / model show?
- What is your scientist famous for?

Ideas for differentiation

Support: Give these learners checklists and clear step-by-step instructions (including completed examples) for every piece of work. Offer suggestions for layout and assist with production of the finished project.

Extension: Give these learners a higher-level, more creative approach to their research. Ask them to consider what implications the scientific discoveries might have for us today and in the future. *How has the work of the scientists influenced current scientific thinking?*

Famous scientist: Sir Isaac Newton

Born: 25 December 1642

Died: 20 March 1727

Facts:

- Sir Isaac Newton was born near Grantham, in Lincolnshire, England and is buried in Westminster Abbey in London.

- He went to Trinity College at Cambridge University in 1661. Whilst he was there he made scientific instruments (equipment), including a telescope. He used his telescope to study the movement of the planets and why they move in their own particular ways. He was a shy man, and so kept his ideas secret for many years.

- In 1665–1666 a terrible disease called the Great Plague swept through England. Isaac was sent home from university when it was closed owing to the plague. He spent two years at home, working on his own ideas and inventions.

- Newton was the first to observe and prove that planets move in elliptical (oval) orbits. He proved this mathematically in a book he wrote called **Principia**.

> **Legend**
>
> The story goes that when an apple fell on his head, he thought of it as like the Moon falling through the sky. This made him think about the pull of the Earth, the Sun and the planets on each other. We now call this 'universal gravitation' – or gravity.

Famous scientist: Albert Einstein

Born: 1879 in Ulm, in Germany

Died: 1955 in USA

Facts:

✔ In 1905 he published his Special Theory of Relativity. Part of this includes his famous equation:

$$E = mc^2$$

E = energy, m = mass and c = the speed of light in a vacuum.

- This equation shows that mass can be changed into energy.
- The Sun gives off energy by converting matter into energy.

✔ In 1907 he began working on his General Theory of Relativity, which he published in 1916. This put forward some new ideas about gravity.

✔ In 1917 he wrote his first paper on cosmology – the study of the universe.

Legend

As a child Einstein asked himself 'What would a beam of light look like if you caught up with it?' (Light travels very fast indeed!)

Famous Chinese scientists

Gan De (also known as Lord Gan)

Facts:

- ✔ Gan De was a Chinese astrologer who studied the stars.
- ✔ He catalogued the stars, along with Shi Shen (see below).
- ✔ He made fairly accurate observations of some of the major planets. He made some of the earliest, detailed observations of Jupiter, saying that it was 'Very large and bright'.
- ✔ He also said that every 12 years it returns to the same position in the sky, and every 370 days it seems to set in the west and rise again 30 days later in the east.

Shi Shen

Facts:

- ✔ Shi Shen worked with Gan De.
- ✔ He found and recorded the positions of 121 stars.
- ✔ He gave us the earliest written recording of sunspots.

Yi Xing

Facts:

- ✔ Yi Xing was also a Chinese astronomer.
- ✔ In the early eighth century he carried out an astronomical survey to help predict solar eclipses.
- ✔ He altered the calendar and made it like the one we have today.
- ✔ He made an accurate working model of the Solar System.

Famous scientist: Aryabhata the Elder

Born: 476 in India

Died: 550

Facts:

- ✔ Aryabhata the Elder wrote at least three books on astronomy and some separate reports that were like poems.

- ✔ In one book he wrote the equivalent of 25 verses about how the planets move. In another section he wrote 50 verses about eclipses.

- ✔ He calculated the circumference (distance all the way round) the Earth and his answer was very close to the number that we agree is true today.

- ✔ He calculated the distances of each planet's orbit and worked out that they move in elliptical (oval) paths.

- ✔ He believed that the planets orbit the Sun because the Earth spins on its own axis.

- ✔ He said that the Moon and planets shine because they reflect the Sun's light.

- ✔ He correctly explained what causes eclipses of both the Sun and the Moon.

Famous scientist: Hipparchus of Rhodes

Born: 190 BCE in Turkey

Died: 120 BCE, probably in Rhodes, Greece

Facts:

- ✔ Hipparchus of Rhodes was an astronomer who studied the stars.
- ✔ He calculated the length of a year to within 6.5 minutes of what we know it to be today.
- ✔ He made a catalogue of 850 stars in 129 BCE.
- ✔ He wrote a book called **Eudoxus**, in which he named and described the constellations – patterns of stars in the sky.
- ✔ He also wrote a long list of bright stars that were always visible. This helped people to tell the time at night.
- ✔ He used a mobile celestial globe, which is a globe that has a map of the sky on it.
- ✔ He studied how the Moon moves by observing it over a long time and using Maths to prove what he thought.
- ✔ He calculated the distance from the Earth to the Moon quite accurately.

Famous scientist: Helen Sharman

Born: 30 May 1963 in Sheffield, England

Facts:

- She works as a chemist in England.
- She was the first British person in space. She was a Project Juno astronaut and visited the MIR space station in 1991. Her mission launched on 18 May 1991.
- She spent 7 days, 21 hours and 13 minutes in space.
- Whilst she was in space, Helen took pictures of the British Isles and spoke to some British schoolchildren via radio.
- She was awarded a medal called the OBE (the Order of the British Empire) in 1993.
- The British School in Assen in The Netherlands is named after her.

Unit 3A: 5.5 Earth's movements

Unit assessment

Questions to ask

- Describe two different ways in which the Earth rotates.
- Why does the Sun **appear** to move across the sky each day?
- How long is a day?
- How many rotations of the Earth does it take to make one day?
- How many days are there in a year?
- Why and how do we have leap years?

Summative assessment activities

Observe the learners while they participate in these activities. You will quickly be able to identify those who appear to be confident and those who may need additional support.

Presentation – research on a space scientist

This activity assesses the learners' understanding and knowledge about the scientist/s they have researched.

You will need:

Finished work from the research project to present to an audience.

What to do

- Ask the learners to present their research findings to a chosen audience – this could be via a poster, leaflet, PowerPoint presentation, model or drama.
- Allow the audience to ask questions at the end.
- Encourage the learners to help create a display using their work on famous space scientists either in your classroom or around school.
- Award marks or comments according to your usual procedures.

Days, weeks and years

This activity assesses the learners' understanding of how the movement of the Earth creates days, weeks and years.

You will need:

A globe, a torch, a large ball to represent the Sun.

What to do

- Ask the learners to use the globe and show you how the Earth rotates on its axis.
- Ask them to describe what one of these rotations results in. (A day and a night.)
- Introduce the Sun and / or the torch and ask the learners to demonstrate when it is day and night during one of the Earth's rotations. How many rotations make a week?
- Ask the learners how we get a year, how many rotations of the Earth on its own axis this takes and how many orbits of the Sun. Check that they know how many weeks there are in a year and how many days in a year.

Written assessment

Distribute photocopiable page 154. The learners should work independently, or with the usual adult support they receive in class.

153

Name: _____

Changing shadows

1. Complete the diagram by drawing the position of the Sun above the horizon at each time of day given.

9:00 a.m.	12:00 noon (midday)
horizon	horizon
3:00 p.m.	6:00 p.m.
horizon	horizon

2. At what time of day will shadows be the shortest? _____

3. Why is this? _____

_____

4. What position is the Sun in the sky when shadows are longest? high / low

Unit 3B: 5.6 Shadows

Light sources 2

Learning objectives

- Observe that shadows are formed when light travelling from a source is blocked. (5Pl1)
- Make relevant observations. (5Eo1)

Resources

Flipchart and markers; a selection of light sources; photocopiable pages 156 and 157.

Starter

- Tell the learners that the work in this unit builds on previous work from Stage 2 units 'Light and dark' and 'Day and night'.
- To establish prior learning, give the learners in small groups a piece of flipchart paper and a marker. Within a set time limit (for example three minutes), ask them to write down as many sources of light as they can think of.
- Share and discuss their lists. Ask them to tell you the light sources that they have identified. Discuss what produces the light for each light source.
  - **All** the learners should be able to identify the following light sources: Sun (burning gas glowing), stars (balls of burning gas that can be smaller or even larger than the Sun – but tend to look smaller as they are further away), lightning (a bright flash of electricity made by a thunderstorm), fire (burning wood or fuel to make flames), candles (wax), torches (cells [batteries]), lightbulbs (electricity).
  - **Most** of the learners should be able to identify the light sources and their source of light (answers in brackets above); these learners may also identify such things as oil lamps (lamp oil).
  - **Some** of the learners may also be able to identify fireflies and some deep-sea fish (have special cells in their bodies that help them to make light).

Main activities

- Discuss which of the above-mentioned sources are natural and which are artificial.
- Give out photocopiable page 156 to the learners who need support and photocopiable page 157 to all the other learners. Explain that they have to identify or name a variety of light sources and (on photocopiable page 157) classify them as natural or artificial.

Plenary

- Discuss the learners' responses to photocopiable pages 156 and 157.
- Ask the learners to make a shadow using a light source of their choice.
- Compare shadows made by an artificial and by a natural light source.

Success criteria

Ask the learners:

- What is the main source of light for us on Earth? (The Sun.)
- Tell me a natural source of light and identify what its source of light is.
- Is a torch a natural or an artificial light source? (Artificial.)
- What is the source of light for a torch? (Cell [battery].)
- How was the shadow formed? (The light from the light source was blocked by an opaque object.)

Ideas for differentiation

Support: Give these learners photocopiable page 156 to complete. Work in a small group with them to clarify their ideas and suggestions.

Extension: Ask these learners to include more detailed information on photocopiable page 157 about what provides the light for each light source they have identified.

Name: _____

Light sources

1. Complete the table to show what the light source is for each object. The first one has been done for you.

2. Now add three more of your own suggestions.

Object	Source of light
torch	a cell (battery)
the Sun	
candle	

Name: _____

Light sources

1. Complete the table to show what the light source is for each object. The first one has been done for you.

2. Then tick (✓) to show which are natural and which are artificial light sources.

3. Now add three more of your own suggestions.

Object	Source of light	Natural?	Artificial?
the Sun	burning gases	✓	
lamp			
oil lamp			

Cambridge Primary: Ready to Go Lessons for Science Stage 5 © Hodder & Stoughton Ltd 2013

Unit 3B: 5.6 Shadows

How shadows are formed

Learning objectives

- Observe that shadows are formed when light travelling from a source is blocked. (5Pl1)
- Make relevant observations. (5Eo1)

Resources

A screen; torches and cells (batteries); a selection of objects for making shadows; photocopiable pages 159 and 160.

Starter

- Set up a screen and play a game where you use a selection of different objects to cast a shadow on the screen, for example a ball, a key, a plant.
- Ask the learners to identify the object that is making the shadow and to explain how they identified it. (It is a solid outline of the object.)
- Allow some of the learners to choose objects to make shadows for the rest of the class to identify.
- Discuss as a class the qualities of shadows. Shadows:
  - are black or dark
  - are the same shape as the object making them
  - can be made bigger or smaller
  - appear to be attached to the object.

Main activities

- Refer back to the Plenary from the previous lesson. Ask the learners to explain why a shadow is formed. Ensure that you use the word 'opaque' and explain that this describes an object that does not allow light to pass through it.
- Discuss how light travels in straight lines. Remind the learners that they might need to recall these facts when trying to explain how shadows are formed.
- Explain to the learners that they are going to choose objects to make shadows with.
- They will then try to make them bigger or smaller or even change shape.

- Give photocopiable page 159 to the learners who need support and explain that they can use this to record how they make their shadows bigger or smaller or change shape.
- Give photocopiable page 160 to all the other learners. Explain that they will draw objects and shadows and identifying some objects from their shadows.

Plenary

- Invite some of the learners who need support to demonstrate the shadows they have made and how they have made them bigger, smaller or change shape.
- Ask questions to guide them when asking them to explain how they think that shadows are formed.
- Invite some of the other learners to draw or make shadows for the rest of the class to guess what is making the shadow.
- Ask these learners to describe how a shadow is made using good scientific language ('opaque' and reference to how light travels).
- Ask any of the learners who have completed the extension activity to show and describe their diagram and to explain it.

Success criteria

Ask the learners:

- What is this a shadow of?
- Why is a shadow black or dark?
- What do we call an object that does not let light pass through it?
- How does light travel?

Ideas for differentiation

Support: Give these learners photocopiable page 159 to complete.

Extension: Ask these learners to draw a diagram of light travelling to make a shadow.

Name: _____

How shadows are formed

You will need:
A torch, an object or toy, a screen or wall.

What to do
- Make a shadow.
- Draw a picture of how you did it.

1. Complete the sentences below.

    a) To make the shadow bigger _____
    _____.

    b) To make the shadow smaller _____
    _____.

2. Did you make the shadow change shape? yes / no

3. Explain in your own words why this happened. _____
_____

Name: _____

How shadows are formed

1. Choose three different objects.

2. Draw each object and the shadow it makes.

Object	Shadow

3. Draw a diagram to show how shadows are made.

4. Use these words to label your diagram.

   light   object   opaque   shadow

Unit 3B: 5.6 Shadows

What makes the sharpest shadow?

Learning objectives

- Observe that shadows are formed when light travelling from a source is blocked. (5Pl1)
- Make relevant observations. (5Eo1)
- Recognise and make predictions from patterns in data and suggest explanations using scientific knowledge and understanding. (5Eo7)

Resources

A selection of transparent, translucent and opaque objects or blocks of material; torches and cells (batteries); A4 card and paper cups to make a screen; photocopiable page 162.

Starter

- Show the learners the selection of objects. In discussion as a class, separate the opaque objects from the rest. Revise that opaque objects do not allow light to pass through them.
- Ask the learners to think with talk partners about how the remaining objects could be classified. After a given time limit (for example two minutes), invite suggestions from the learners.
- Introduce the words 'transparent' and 'translucent'. Do not give definitions of these two kinds of materials – this is what the learners will find out in this lesson.

Main activities

- Show the learners how to make a screen from a piece of A4 card and two paper cups. Pre-prepare two paper cups with a slit cut in the base for a piece of A4 card to slot into. This makes a well-supported base and screen for the purposes of this experiment.
- Organise the learners into groups – either ability groups or mixed-ability groups, as you prefer. Ensure that each group is given at least one opaque, one transparent and one translucent object or piece of material.
- Give out photocopiable page 162 to all the learners.
- Explain that they have to make shadows using the three different kinds of material and compare them.

Plenary

- Invite different groups of learners to demonstrate the shadows they have made during the course of the lesson.
- Ask them to comment on the differences in the types of shadow produced by the transparent, translucent and opaque materials. Are the shadows sharp or fuzzy?
- Define the differences between opaque (doesn't allow any light to pass through it), transparent (allows light to pass through it) and translucent (allows some light to pass through it).

Success criteria

Ask the learners:

- Which material made the most distinct shadow?
- Which material made the least effective shadow?
- What do we call materials that do not allow light to pass through them?
- How would you describe a transparent material?
- How much light does a translucent material allow to pass through it?

Ideas for differentiation

Support: Either allow these learners to work in a mixed-ability group, or assist them in completing photocopiable page 162.

Extension: Ask these learners to try different opaque materials and rank them according to which makes the sharpest shadow.

Name: _____

What makes the sharpest shadow?

You will need:

A torch and cells (batteries), a piece of A4 card, two paper cups with slits in them, something opaque, something transparent, something translucent.

What to do

- Make a screen from the A4 card and paper cups.
- Use the torch to try to make a shadow using the opaque object or material.
- Repeat the activity using the transparent and translucent objects or materials.
- Compare the different shadows.

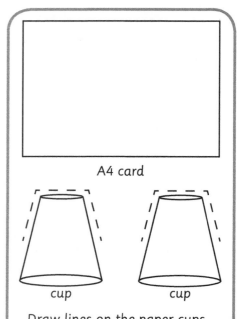

A4 card

cup   cup

Draw lines on the paper cups and cut slits in the bases.

Slide the A4 card into the slits to make a screen.

Results

1. Complete the table to show which objects or materials you used each time and what each shadow looked like.

Opaque	Transparent	Translucent
Name of object or material	Name of object or material	Name of object or material
Shadow	Shadow	Shadow

2. Which object or material made the sharpest shadow and why?

_____

_____

Unit 3B: 5.6 Shadows

Changing the size of shadows

Learning objectives

- Investigate how the size of a shadow is affected by the position of the object. (5Pl2)
- Measure volume, temperature, time, length and force. (5Eo2)

Resources

A large space outside in the playground on a sunny day; hats; sun cream; a selection of torches and cells (batteries); A4 card; paper cups with slits in them; a selection of small objects for making shadows; graph paper; rulers; photocopiable pages 164 and 165.

Starter

- Go outside and ask the learners to stand in a place where they can see their shadow. Ensure that they are wearing suitable footwear for running around in and a hat and sun cream if necessary.
- Play a game of shadow 'tag'. Choose one learner to be 'on'. This person has to chase the other learners around and stand on their shadow. When this happens, the learner has to stand out of the game. Choose an appropriate place for these learners to congregate when this happens. The winner is the last person whose shadow is trodden on by the learner who is 'on'.
- An alternative game is to choose several learners to be 'on'. Give these learners a coloured PE bib or tape to wear as a marker for identification.
- Another alternative is to have several learners as 'on' but also choose another learner and give them a different colour to wear. This time the learners freeze when their shadow is trodden on. Part way into the game, allow the learner with the different colour to join the game. This learner can 'free' the frozen learners and the game takes longer.

Main activities

- Return inside and discuss what the learners noticed about their shadows. Some will still think that their shadow is 'stuck' to their feet. Ask them to think about when or if they noticed how their shadow changed shape during the game.
- Explain that in this lesson they will make shadows and think about what makes a difference to their size.
- Give out photocopiable page 164 and explain that the learners need to follow the instructions and complete the page to show what they have found out about changing the size of shadows.

Plenary

- Invite different individual or groups of learners to demonstrate what they have done.
- Discuss how to make a shadow bigger and smaller.

Success criteria

Ask the learners:

- How did you make the shadow bigger?
- How did you make the shadow smaller?
- Why did this happen?

Ideas for differentiation

Support: Either work with these learners together in a small group, or allow them to work in mixed-ability groups.

Extension: Challenge these learners to make the smallest and biggest shadows of their object that they can, or give them photocopiable page 165 to complete after photocopiable page 164.

163

Name: _____

Changing shadows

You will need:
A torch and cells (batteries), a piece of A4 card, two paper cups with slits in them, a small object to make a shadow with, a ruler, graph paper to sit the object on.

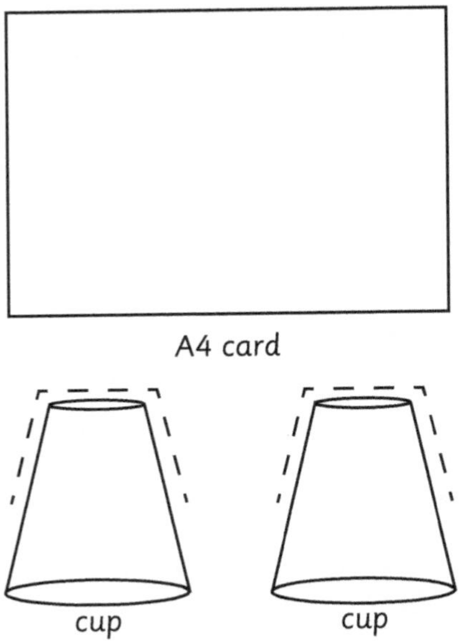

What to do
- Make a screen from the A4 card and paper cups.
- Place the graph paper between the torch and the screen.
- Place the object on the graph paper, then record the distance from the torch and the height of the shadow in the table below.
- Repeat the activity, placing the object at different distances each time.

A4 card

cup   cup

Draw lines on the paper cups and cut slits in the bases.

Slide the A4 card into the slits to make a screen.

Results

1. Write your results in the table.

Distance from torch in cm	Height of shadow in cm

2. How does the shadow change?

   a) The shadow is bigger when _____.

   b) The shadow is longer when _____.

Name: _____

Altering shadows

1. Use your results from the 'Changing shadows' activity to draw a line graph.

2. Remember to:
   - include a title
   - label both axes
   - choose a suitable scale
   - use a ruler.

Title: Graph to show _____

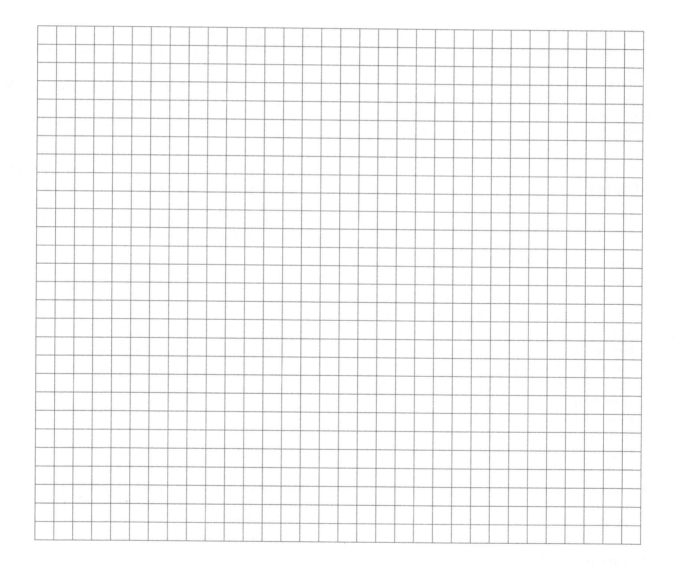

3. What do you notice about the height of the shadow as the distance increases?

_____

_____

Unit 3B: 5.6 Shadows

Investigating shadows

Learning objectives

- Investigate how the size of a shadow is affected by the position of the object. (5Pl2)
- Make relevant observations. (5Eo1)
- Decide whether results support predictions. (5Eo5)

Resources

Outdoor space on a sunny day; hats; sun cream; a digital camera (if available); torches and cells (batteries); a selection of translucent and opaque objects or blocks of material; a screen; A4 card; paper cups with slits in for screen supports; A4 paper; sticky tack; scissors; glue sticks; photocopiable pages 167, 168 and 169.

Starter

- Take the learners outside on a sunny day and ask them to make the longest shadow that they can. Ensure that they are wearing suitable footwear and have adequate sun protection.
- Look at and discuss the various ways in which different learners manage to achieve this. Take some photographs, if appropriate. Allow the learners to take photographs also.
- Repeat the activity, this time asking the learners to make as short a shadow as possible. Again, observe and discuss their ideas, taking photographs as necessary.
- Now ask them to try to get rid of their shadow. If any of them manages to do this, notice exactly how it is achieved (they will have to go into the shade).
- Go back inside and give out photocopiable page 167 for the learners to record how they did the activities.

Main activities

- Now arrange the learners in groups of three and ask them to investigate making the longest and shortest shadow for some objects and how to make shadows disappear.

- Give out photocopiable pages 168 and 169 for them to complete and show their method. The group members will each have to choose a different object to use for photocopiable page 169.

Plenary

- Invite different learners to share their conclusions with the rest of the class.
- Discuss their findings and correct any misconceptions arising.
- Ensure that all the learners know that the closer to the torch (or light source) an object is, the bigger the shadow will be.
- Also emphasise that the opposite is true, that is, that the further away from the torch (or light source) an object is, the smaller the shadow will be.
- Demonstrate and explain how light travels in straight lines and that the rays are blocked by the opaque or translucent object and, therefore, do not pass through the object but create a dark shadow behind it – on the opposite side to the light source.

Success criteria

Ask the learners:

- What happened each time?
- Who created the longest shadow?
- Who made the shortest shadow?
- Why does this happen?

Ideas for differentiation

Support: Assist these learners with drawing around the shadows produced on the screen and cutting them out to use on photocopiable page 169.

Extension: Ask these learners to draw a diagram showing how light travels to illustrate what they have observed in this lesson.

Name: _____

Changing my own shadow

1. Draw what you did to make your longest shadow.

2. Describe how you made the shortest shadow possible.

3. Explain how it was possible to hide from your shadow.

_____
_____

Name: _____

Investigating shadows 1

You will need:
A torch and cells (batteries), A4 card, two paper cups with slits in, at least three different objects, A4 paper, sticky tack, a pencil.

What to do

- Make a screen from the A4 card and paper cups. Fix the A4 paper to the screen.
- Predict which object will make the longest and which will make the shortest shadow.

Longest = _____   Shortest = _____

- Take one object and make the longest shadow you can and draw around it.
- Repeat the activity to make the smallest shadow.
- Do this for at least two more objects.
- Draw the objects you used below.

Name: _____

Investigating shadows 2

1. Choose one of your objects.

2. Cut out and stick the shadows you made on this page.

Shortest shadow

```
┌─────────────────────────────────────────┐
│                                         │
│                                         │
│                                         │
│                                         │
└─────────────────────────────────────────┘
```

Longest shadow (you may need to use the back of this page)

```
┌─────────────────────────────────────────┐
│                                         │
│                                         │
│                                         │
│                                         │
└─────────────────────────────────────────┘
```

Conclusion (what you found out)

3. Complete these sentences:

   a) When the object is close to the torch, the shadow is _____.

   b) When the object is further away from the torch the shadow is _____.

Unit 3B: 5.6 Shadows

What affects the size of a shadow?

Learning objectives

- Investigate how the size of a shadow is affected by the position of the object. (5Pl2)
- Use observation and measurement to test predictions and make links. (5Ep2)
- Make relevant observations. (5Eo1)

Resources

Torches and cells (batteries); a selection of translucent and opaque materials; a selection of small objects for making shadows; A4 card; paper cups with slits in for screen bases; photocopiable pages 171 and 172.

Starter

- Make a screen by slotting the A4 card into the two paper cups. Set up a shadow on the screen. Invite different learners to come forward and either make the shadow bigger or smaller.
- Discuss why this happens – this is to recap from the previous lesson's Plenary. The closer the object is to the torch (light source) the bigger the shadow; the further away from the torch (light source), the smaller the shadow.
- Demonstrate this by drawing a diagram from a bird's-eye view of how the shadows are formed (see photocopiable page 189). Include the light rays travelling in straight lines away from the torch, but not being able to penetrate the opaque object, so that a shadow forms behind the object where no light can pass.

Main activities

- Explain that in today's lesson the learners are going to find out if changing the position of the light source makes a difference to the appearance, position and size of the shadow produced.

- Demonstrate what the learners have to do with your torch switched off. Explain that they will need to do it with their torches switched on. Hold the torch on one side of the object, low down. Then hold the torch on the same side of the object and the same distance away from it, but hold the torch higher up. Finally, hold the torch directly above the object.
- In class discussion, predict what might happen in each instance. How do the learners know this? (Some might relate it to the position of the Sun in the sky – acknowledge this, but do not elaborate on the point. This will be the focus of the next lesson.)
- Give out photocopiable page 171 and explain what the learners have to do to complete it – draw a diagram and then record the differences in shadows produced with the torch in different positions.

Plenary

- Discuss the answers to the sentences at the bottom of photocopiable page 171.

Success criteria

Ask the learners:

- How does light travel?
- Why does a shadow form?
- When the light source is low, what does the shadow look like?
- When the light source is at its highest, what is the shadow like then?

Ideas for differentiation

Support: Assist these learners in completing photocopiable page 171.

Extension: Give these learners photocopiable page 172 to complete after they have completed photocopiable page 171.

Name: _____

What affects the size of a shadow?

1. Draw light rays on the diagram to show how the shadow is formed.

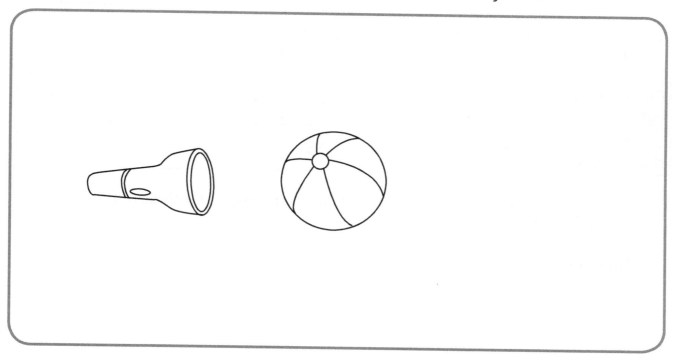

2. Draw the shadows made with the torch in each of the three different positions.

Torch low down	Torch higher up	Torch directly above

3. Use these words to complete the sentences.

> higher   longer   lower   shorter

a) The _____ the light source the _____ the shadow.

b) The _____ the light source the _____ the shadow.

Cambridge Primary: Ready to Go Lessons for Science Stage 5 © Hodder & Stoughton Ltd 2013

Name: _____

Shadow graph

Here are some results from an experiment by some Stage 5 schoolchildren.

1. Complete the table to predict the results for the height of the shadow at 60, 70 and 80 cm distances away from the light.

Distance from light in cm	10	20	30	40	50	60	70	80
Length of shadow in cm	50	45	40	35	30			

2. Use these results to draw a line graph below. Remember to include a title and to label the axes.

   Title: Graph to show _____

3. How could you change the position of the light source to make the shadows as short as possible?

   _____

   _____

Unit 3B: 5.6 Shadows

How do shadows change throughout the day?

Learning objectives

- Observe that shadows change in length and position throughout the day. (5Pl3)
- Use observation and measurement to test predictions and make links. (5Ep2)
- Decide whether results support predictions. (5Eo5)

Resources

Outdoor space where the Sun will shine throughout the school day; rulers or metre sticks; buckets of sand or soil to hold the stick upright; chalk; tape measures; photocopiable pages 174, 175 and 176.

Starter

- Go outside. Ask the learners in pairs or small groups to choose a good position to place a ruler or metre stick in a bucket of sand or soil and measure the length of the shadow. Ask them to record the measurement on photocopiable page 174.
- Ask them to predict what the length of the shadow will be one hour later and to record the prediction on photocopiable page 174.
- Go back inside the classroom. Share measurements – are they similar or very different? Ask the learners what they have predicted, and listen to their reasons why. Some of them may relate their predictions to the position of the Sun in the sky.

Main activities

- Make sure that all the learners know that the Sun is the outdoor source of light and that it is a natural light source.
- Ask the learners to think back to the previous lesson. Recall what happened when the position of the torch was changed – the lower the light source, the longer the shadow, and vice versa.
- Use photocopiable page 174 for the learners to record their stick's shadow length at hourly intervals throughout the school day. Ask them to mark the shadow in chalk on the playground, measure it and write the measurement by the side.

- After each new measurement is taken, give the learners time to think about the changes in length and the position of the shadow.
- With talk partners, ask the learners to consider what has changed regarding the length of the shadow – is it longer or shorter than before? How has the shadow moved?
- Consider the learners' explanations for why this might be – do not comment on these observations until you do the Plenary, after all the results have been recorded.

Plenary

- Compare measurements and help the learners to interpret their measurements by asking pertinent questions (see 'Success criteria').
- Relate the changes in position and size of the shadows each time with the position of the Sun in the sky.
- Some of the learners will still think that the Sun rises and sets. Explain that the Sun only appears to move because the Earth is spinning on its axis. Make sure that they all understand that the Sun does not move.
- Give out photocopiable page 175 to check their understanding.

Success criteria

Ask the learners:

- At what time was your stick's shadow the shortest?
- Where is the Sun in the sky at this time?
- At what time did you see the longest shadow?
- Why did the shadows change position and length throughout the day?

Ideas for differentiation

Support: Allow these learners to work in a mixed-ability group, or work with them in a small group.

Extension: Give these learners photocopiable page 176 to complete, in addition to photocopiable pages 174 and 175.

Name: _____

Shadow sticks 1

Record your results in the table below.

Time	Position of shadow (draw)	Length of shadow

Name: _____

Shadow sticks 2

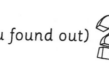

Conclusion (what you found out)

1. Match the beginning and end of the sentences.
   Join them with a line.

The shadow is short when	at its highest in the sky.
At midday, the Sun is	it is midday.
The longest shadows are	does not move.
The Sun	when the Sun is low in the sky.

2. Why does the Sun appear to move across the sky every day?

_____
_____
_____
_____

Cambridge Primary: Ready to Go Lessons for Science Stage 5 © Hodder & Stoughton Ltd 2013

Name: _____

Shadows throughout the day

Draw the shadow of the tree at the time of day shown in each box.

6:00 a.m.	12 noon (midday)
horizon	horizon

3:00 p.m.	6:00 p.m.
horizon	horizon

Unit 3B: 5.6 Shadows

Shadows during the course of a day

Learning objectives

- Observe that shadows change in length and position throughout the day. (5Pl3)
- Use observation and measurement to test predictions and make links. (5Ep2)
- Make relevant observations. (5Eo1)

Resources

Sticky tack; a window in the Sun all day; a globe; a torch and cells (batteries); an object to make a shadow; a large ball; photocopiable page 178; a calendar or diary with local sunrise and sunset times on it.

Starter

- Ask the learners to discuss with talk partners the changes you observed last lesson in terms of length and position of shadows throughout the day.
- Remind the learners that you mimicked the Sun's movements during the day when you did the experiment on page 170 and moved the torch from low to high, then lower again. Relate this to how the Sun appears to travel across the sky daily.
- During the course of the day, place a ball of sticky tack on a window through which the Sun can be seen all day. Mark the position of the Sun as viewed through the window. Repeat this hourly throughout the day, using a new ball of sticky tack each time. (This was done as the Starter activity on page 120.)
- Think about what pattern the learners might expect to see during the course of a school day. (A curved pattern should begin to emerge.)

Main activities

- Using the globe, demonstrate, or ask the learners to demonstrate, the Earth's rotation. Use a torch to represent the Sun, or a much bigger sphere than the globe to remind the learners that the Sun does not move and that it is only because of the Earth spinning that the Sun **appears** to move across the sky daily.

- Recap the position of the Sun in the sky in relation to the horizon when it is sunrise. Some of the learners will need to have the word 'horizon' explained to them (the point where the land and sky appear to meet in the distance, as far away as you can see).
- Then consider the position of the Sun in the sky at noon (midday). This is when the Sun is at its highest or its zenith. Refer to the shadow-stick experiment results from the previous lesson to compare what the length of the shadows was at midday.
- Finally consider the position of the Sun in relation to the horizon as sunset approaches.
- Give out photocopiable page 178 to all the learners except those who need extension activities. Explain that they have to draw pictures of a shadow throughout the day.

Plenary

- Check that the learners remember and understand what exactly is happening during the Sun's apparent daily movements.

Success criteria

Ask the learners:

- Which direction does the Sun appear to rise from?
- Why are shadows shortest at midday (noon)?
- In which direction on the horizon does the Sun appear to set?
- How does the Earth move during 24 hours?

Ideas for differentiation

Support: Assist these learners with completing photocopiable page 178.

Extension: Ask these learners to make a table of sunrise and sunset times locally for the 21st of the month during a year. What do they notice about 21 June and 21 December?

Name: _____

Shadows throughout the day

1. Draw a stick person on the horizon in the middle of the line.

2. Look at the time in each box and draw the shadow made by the Sun. Remember it changes length and position.

6:00 a.m.

_____ horizon

12:00 noon

horizon _____

6:00 p.m.

horizon _____

Unit 3B: 5.6 Shadows

Measuring light intensity

Learning objectives

- Know that light intensity can be measured. (5Pl4)
- Discuss the need for repeated observations and measurements. (5Eo3)
- Interpret data and think about whether it is sufficient to draw conclusions. (5Eo8)

Resources

A light meter for a camera; a light meter for use in a laboratory; a digital camera with a light meter that can be altered manually as well as automatically; photographs from the internet or books if the equipment is not available; photocopiable page 180; calculators.

Starter

- Ask the learners if they know what piece of equipment helps us to measure light (a light meter).
- Ask the learners to discuss with talk partners which kinds of jobs require that the intensity of light is measured. (Their most common suggestion might be photographers.)
- Describe the intensity of light as how bright it is. Tell the learners that sometimes light meters are inside a camera, but sometimes they can be attached to a camera externally.
- Light meters work by having a special electric cell inside them that reacts to light. When this happens, an electric current is produced. The current makes a pointer on a dial move or some numbers on the screen of the light meter appear.

Main activities

- If light meters are available, the learners can use them to record light intensities in different situations.
- In groups or pairs, ask them to discuss a place where there is a lot of light and a place where there is less light. Persuade them to think about a bright place and a dark place for comparison. These will be the places where the learners can try using the light meter.
- Either demonstrate (if only one is available) or allow the learners time to become accustomed to using the light meters. Take them through step by step, especially in recording the results and what they mean.
- Give out photocopiable page 180 for them to record their readings on.

Plenary

- Check the learners' calculations of average light intensities.
- Discuss their findings.
- Explain that light meters can be used at a lower secondary level to compare different amounts of light coming from different types of lamps (lightbulbs).

Success criteria

Ask the learners:

- What piece of equipment can we use to measure light intensity?
- Which professionals might use light meters in their daily work?
- Why is it sometimes useful to take repeat measurements and then take an average result? (It makes the data more reliable.)
- What is the difference between the lowest and highest light intensities that you measured?

Ideas for differentiation

Support: These learners will need close supervision in using the equipment and taking readings.

Extension: Ask these learners to find out how underwater cameramen use light meters to take good photographs. Ask them to create a poster to display and show the rest of the class.

Name: _____

Measuring light intensity

1. Record your results in the table. Use a calculator to help you to work out the averages if you need to.

Place	Reading 1	Reading 2	Reading 3	Average reading
light place =				
shadier place =				

2. Why is it sometimes useful to take average readings for results?
   _____

3. What was the highest light intensity that you measured?
   _____

4. What is the difference between the lowest average light intensity and the highest average light intensity measurements? _____

Show your calculation.

Unit 3B: 5.6 Shadows

How does light travel through some materials?

Learning objectives

- Explore how opaque materials do not let light through and transparent materials let a lot of light through. (5Pl5)
- Make predictions of what will happen based on scientific knowledge and understanding, and suggest and communicate how to test these. (5Ep3)
- Identify factors that need to be taken into account in different contexts. (5Ep6)

Resources

A selection of opaque, transparent and translucent materials or objects; torches and cells (batteries); A4 card; paper cups with slits in for screen bases; a set of blocks of different materials, e.g. wood, glass, metal, perspex; plastic bottles; sandpaper; black paper or paint; greaseproof paper; glass beakers; water; fruit squash concentrate that needs diluting to make a drink; milk; photocopiable pages 182–187.

Starter

- Ask the learners to discuss with talk partners how and why objects are easy or difficult to identify from their shadows. (They are easy to identify if the outline is clear and it is a distinctive object, more difficult if an item is a regular, solid shape or the object is viewed from different angles.)
- Ask the learners to discuss with talk partners which materials or objects make good shadows. Recap on the qualities of opaque, transparent and translucent materials.

Main activities

- Organise the learners into ability groups and give each group one of the following questions:
  - The learners who need support: *Which type of material makes the best shadow?* (This is revision from Stage 2.) Give out photocopiable pages 182 and 183 to these learners.
  - The average-ability learners: *Which plastic bottle makes the best shadow?* Give these learners photocopiable pages 184 and 185.
  - The learners who need extension: *Which liquid makes the best shadow?* Give these learners photocopiable pages 186 and 187.
- Allow the learners to plan and carry out their investigation, completing the relevant photocopiable pages as they do so.
- The average-ability learners will need help in deciding how to make a translucent plastic bottle (use sandpaper over its whole surface or cover it with greaseproof paper). They may decide to paint or cover a bottle to make it opaque – or even fill it with sand. Allow them to make choices, but steer them to using resources that are readily available in the classroom.
- The learners who need extension may ask how to get a translucent liquid – let them dilute some squash, or they might suggest other liquids, such as watery paint.

Plenary

- Ask for answers to the questions posed and a demonstration from different groups to show what they did.
- Confirm that opaque materials do not let light pass through them, translucent materials let some light pass through and transparent materials let all light pass through.

Success criteria

Ask the learners:

- What type of material makes the best shadow?
- How can you make a plastic bottle translucent or opaque?
- Which opaque liquid did you use?
- Why do opaque objects or materials produce the clearest shadows?

Ideas for differentiation

Support: Give these learners blocks of materials to work with.

Extension: Give these learners liquids to investigate.

Name: _____

Which material makes the best shadow? 1

You will need:
Blocks of different materials (for example wood, metal, plastic, glass), a torch and cells (batteries), A4 card and paper cups to make a screen.

Method (what you did)

1. Draw or write to show how you set up the equipment.

Results (what happened?)

2. Complete the table to show what happened each time.

Shadow (draw and shade)	Shadow (draw and shade)
Material _____	Material _____
Material _____	Material _____

182   Cambridge Primary: Ready to Go Lessons for Science Stage 5 © Hodder & Stoughton Ltd 2013

Name: _____

Which material makes the best shadow? 2

Results (continued)

3. Complete these sentences:

   a) The _____ made the darkest shadow.

   b) The _____ made a faint shadow.

   c) The _____ made no shadow at all.

4. List the materials you are using and tick (✓) to show if they are opaque, transparent or translucent.

Material	Opaque? (✓)	Transparent? (✓)	Translucent? (✓)

Conclusion (what you found out)

5. Complete the sentences below using these words.

   opaque   transparent   translucent

   a) Good shadows are made when the material is _____ because these materials do not let any light through.

   b) Faint shadows are cast by _____ materials, which let some light through.

   c) _____ materials let a lot of light through and so do not make a shadow.

Cambridge Primary: Ready to Go Lessons for Science Stage 5 © Hodder & Stoughton Ltd 2013

Name: _____

Which plastic bottle makes the best shadow? 1

You will need:

Three plastic bottles, a torch and cells (batteries), A4 card and paper cups to make a screen.

1. How will you make one of the bottles translucent?

   _____

   _____

2. How will you make another bottle opaque?

   _____

   _____

What to do

- Make shadows using each bottle you have prepared.
- Draw a diagram below to show how you set it up.

Name: _____

Which plastic bottle makes the best shadow? 2

3. How will you make this a fair test? (What factors will you keep the same?)

- _____
- _____
- _____
- _____

4. Which factor will you change each time?

_____

Results (what happened?)

```
┌─────────────────────────────────────────────┐
│                                             │
│                                             │
│                                             │
│                                             │
│                                             │
└─────────────────────────────────────────────┘
```

5. Which bottle made the best shadow and why?

_____
_____
_____

Cambridge Primary: Ready to Go Lessons for Science Stage 5 © Hodder & Stoughton Ltd 2013

Name: _____

Which liquid made the best shadow? 1

You will need:
A transparent liquid, a translucent liquid, an opaque liquid, three glass beakers, a torch and cells (batteries), A4 card and paper cups to make a screen.

1. Record the names of the liquids you chose to use in the table below.

	Name of liquid
Transparent	
Translucent	
Opaque	

Method (what you did)

2. Draw or write to show how you set it up.

Name: _____

Which liquid made the best shadow? 2

Results (what happened?)

Conclusion (what you found out)

3. Which liquid made the best shadow? _____

4. Explain why:

   _____

   _____

   _____

Cambridge Primary: Ready to Go Lessons for Science Stage 5 © Hodder & Stoughton Ltd 2013

Unit 3B: 5.6 Shadows

Fun with shadows!

Learning objectives

- Explore how opaque materials do not let light through and transparent materials let a lot of light through. (5Pl5)
- Make relevant observations. (5Eo1)

Resources

A selection of transparent, translucent and opaque objects; torches and cells (batteries); screens; black paper; card; scissors; sticky tape; small sticks; digital camera or big sheets of white paper; black paint; painting equipment; photocopiable pages 189 and 190.

Starter

- In pairs or small groups, organise a hunt around the classroom to find an example of an opaque, a transparent and a translucent object. (This will depend on the availability of objects.)
- Alternatively, prepare sets of labels containing the words 'opaque', 'transparent' and 'translucent' and choose individual learners to label objects around the room accordingly.
- Revise that opaque objects do not let light through, transparent ones do and translucent objects let some light through.

Main activities

- Ask the learners to think about why these groups of materials (opaque, transparent and translucent) act as they do.
- Revise what happens when light hits these objects, using an enlarged version of photocopiable page 189.
- Discuss examples of each type of material and consider how light rays are affected by it when they hit the object.
- Give out photocopiable page 190 to all the learners. Go through the instructions, demonstrating each step in the making of a shadow puppet.
- Allow the learners to make a shadow puppet each and perhaps to work with a group of friends to produce a short puppet-show to show to the rest of the class or to younger learners.
- Show the learners who need extension how to make silhouettes: Take a photograph in profile view. Print it out. Cut around the head outline. Paint the outline black and stick it on a white background.
- Alternatively, project the learner in profile onto a large sheet of white paper on a screen. Draw around the outline. Cut it out, paint it black and stick it on a white background.

Plenary

- Invite the learners who need support to identify objects as opaque, transparent or translucent, as in the Starter activity.
- Invite the average-ability learners to try to explain what happens to light rays when they hit each of these objects.
- Watch some shadow-puppet performances from different groups.

Success criteria

Ask the learners:

- Is this opaque, transparent or translucent?
- Why do we get a shadow when light rays hit a solid object?
- How much light does a transparent object let through, compared to a translucent object?
- How can you make it look as if your shadow puppet is going away into the distance?
- Who is this a silhouette of?

Ideas for differentiation

Support: Assist these learners with cutting skills and in following the instructions to make a shadow puppet.

Extension: Show these learners how to make silhouettes.

Looking at light

The diagrams below show what happens to light when it hits a transparent, translucent or opaque object.

Transparent:
light travels straight through

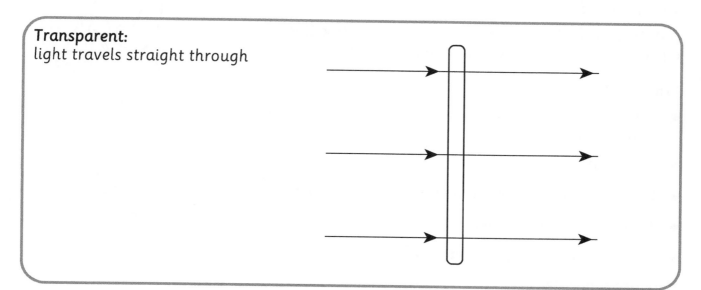

Translucent:
light passes through but is scattered in random directions

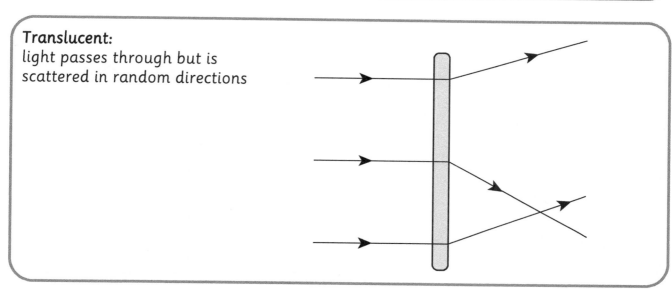

Opaque:
light does not pass through – it is either reflected or absorbed

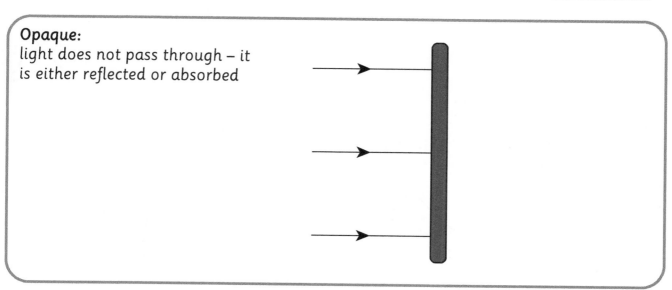

Cambridge Primary: Ready to Go Lessons for Science Stage 5 © Hodder & Stoughton Ltd 2013

Name: _____

Shadow puppets

You will need:
A pencil, card, sticky tape, scissors, a small stick (for example a pea-stick), a screen and strong light source.

What to do
- Draw the outline only of your character – a person, an animal or an object.
- Cut it out.
- Use sticky tape to attach the pea-stick to the back of it.
- Try using it on the screen.
- Can you make it bigger and smaller?

Here are some ideas.

Unit 3B: 5.6 Shadows

Unit assessment

Questions to ask

- What is a light source?
- How are shadows formed?
- What is the difference between objects that are opaque, transparent and translucent?
- How can you change the size of a shadow?
- What equipment do we use to measure light intensity?

Summative assessment activities

Observe the learners while they participate in these activities. You will quickly be able to identify those who appear to be confident and those who may need additional support.

Light sources

This activity assesses the learners' understanding of light sources.

You will need:

A selection of natural and artificial light sources – actual or pictures, or a combination.

What to do

- Ask the learners individually to identify a natural and an artificial light source and to identify its light source.
- Ask them to name and describe at least two more.
- Record their responses on a class checklist as an indicator of their knowledge about light sources.

What makes the shadow?

This activity assesses the learners' understanding of shadows.

You will need:

A set of pre-prepared cards containing pictures of objects and a corresponding card showing the shadow created by that object, for example a key, a teddy bear, a book, a ruler. Include some objects where shadows are created from a less-obvious angle to make it more difficult for the learners to identify them.

What to do

- Ask the learners to match each object card to its shadow card.
- Alternatively play this as a game where the learners take turns and the winner of the game is the learner with the most pairs of cards.
- Discuss how easy or difficult it was to match the pairs and why.

Written assessment

Distribute photocopiable page 192. The learners should work independently, or with the usual adult support they receive in class.

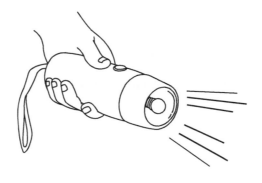

Name: _____

Opaque, transparent or translucent?

1. Complete the table below.

Diagram	Opaque, transparent or translucent?	Name of an object that fits this description
(light rays passing straight through)		
(light rays scattered through)		
(light rays blocked)		

2. Why do opaque objects make the sharpest shadows?

_____

_____